Problem Solving through Investigation

- Book 1

George Moore

World Teachers Press

Published with the permission of R.I.C. Publications Pty. Ltd.

Copyright © 1997 by Didax, Inc., Rowley, MA 01969. All rights reserved.

First published by R.I.C. Publications Pty. Ltd., Perth, Western Australia.

Limited reproduction permission: The publisher grants permission to individual teachers who have purchased this book to reproduce the blackline masters as needed for use with their own students. Reproduction for an entire school or school district or for commercial use is prohibited.

Printed in the United States of America.

Order Number 2-5026
ISBN 1-885111-39-8

    C D E F   99 00

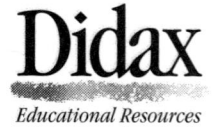

395 Main Street
Rowley, MA 01969

Foreword

Problem Solving Through Investigation is a two book set of enrichment activities for grades five through ten. Book 1 is intended for grades five to eight and Book 2 is intended for grades seven to ten.

Each book contains over 40 separate activity sheets that reinforce concepts in number, measurement and space. These activities have been designed to stimulate interest in mathematics, and can be used in the class or for assignments. Many of the exercises are suitable for partner or group work, providing valuable interaction between students, a desirable feature of math investigation activities.

The activities in each book are both fundamental in concept and challenging in application. They address interesting and stimulating areas of mathematics that will assist with the development of positive attitudes.

Contents

Teacher Information 4-5

Number
Factors 6
Factor Patterns 7
Products and Isometric Shapes 8
Odd Numbers 9
Prime Numbers 10
Composite Numbers 11
Prime and Composite Numbers 12
Commutative Law 13
Square and Triangular Numbers 14
Probability 15
Magic Squares 16
Fun with the Magical 9! 17
Operations — Calculator Work 18
Number Patterns (1) 19
Number Patterns (2) 20
Number Patterns (3) 21
Number Systems 22
Index Notation — Calculator Work 23

Measurement
Mapping 24
Perimeter 25
Perimeter and Area 26
Triangles 27
Angles 28
Decades and Centuries 29
Polygons 30

Space
Duplation 31
Conservation of Area 32
Congruency 33
Reflective Symmetry 34
Closed Curves 35
The Circle 36
Coordinates (1) 37
Coordinates (2) 38
Area 39
Line Segments 40
Coordinates (3) 41
Rigid Shapes 42

Answers 43-45

Teacher Information

Problem Solving Through Investigation is a
two book series that provides activities to challenge and motivate young minds. Each activity is provided with the specific intent to provide a problem that requires the use of a variety of problem solving strategies and skills.

Activites are drawn from the number, space and measurement areas of mathematics to provide a variety of content and interest.

The activities in this book can be used in a variety of class situations including:

1. Whole class;
2. Small group;
3. Individual;
4. Homework; or
5. Activity Center.

Answers are provided but should not be consulted to solve problems but rather confirm the answers that are obtained.

Measurement
As an added problem solving component the measurement activities in this book have been provided in metric format. As well as providing an added challenge this will provide valuable experience with the metric measurement system.

Problem-Solving Strategies

Students should be encouraged to use a variety of problem-solving strategies to solve the problems in this book. These include:
- exploring patterns;
- organizing information;
- finding all answers; and
- using manipulating.

More information on these strategies can be found in the World Teachers Press series 'Problem Solving with Math'.

The following points should be emphasized when introducing the activities:

1. **Comprehend the Problem**
   It is vital that students have a clear understanding of what the problem is asking.

2. **Select Appropriate Activities**
   After identifying the problem the student needs to select the strategy or strategies that are most appropriate (see above). Students should be encouraged to change their strategy if success is not forthcoming. Discussion with partners or small groups can be very effective in this situation.

3. **Review the Process**
   After solving or attempting to solve the problem the process used should be reviewed. Questions such as the following will provide guidelines for this:
   (i) Did I identify the problem adequately?;
   (ii) Were the strategies used the most efficient/effective?; and
   (iii) What would I do differently?

Teacher Information – Example Lesson Development

The following is an example lesson development using one of the pages in this book. It demonstrates how the activity could be introduced, developed and extended.

Activity   Factors: page 6

Introductory Work

This activity can be linked to other aspects of the math curriculum, including fractions and division. While a problem-solving activity, it reinforces the basic knowledge of factors.

Discussion can be based on what a factor is and why we need to know the factors of numbers.

Using the Worksheet

An ideal activity for group work. This activity can use multi-based cubes as well as graph paper. Some students will be capable of working independently through the entire activity.

Emphasize the meaning of factors so students understand what they are looking for.

Complete the first examples with the group using paper and/or cubes.

Students work independently through the examples provided using paper and/or cubes.

An opportunity exists to identify and name 'prime numbers'.

Extension

Provide more difficult examples to be completed and encourage students to calculate factors mentally.

World Teachers Press — Problem Solving Through Investigation – Book 1

Factors

A *factor* is a number which will divide *exactly* into another number.
For example: the factors of 21 are 1, 3, 7, 21.

Here is a different way of finding the factors of **16**. We can arrange 16 units into only three different rectangles.
Remember: a square is a special rectangle with congruent sides.
The *factors* of 16 are 1, 2, 4, 8 and 16.
16 has more than two factors so it is a **composite number**.

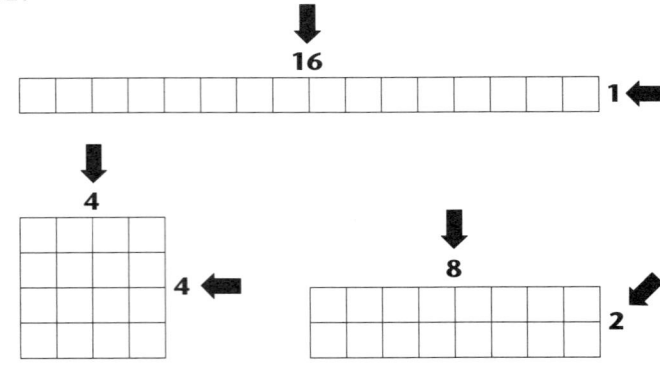

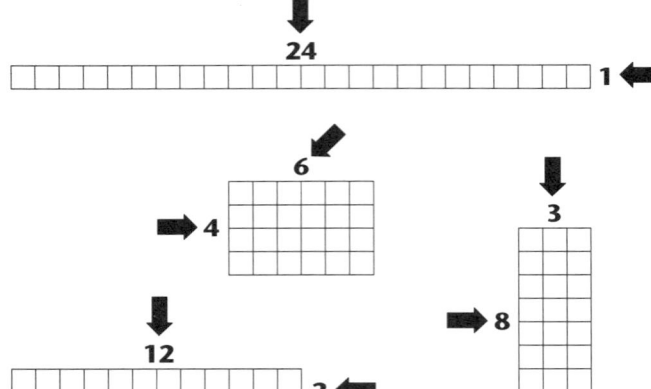

We can arrange **24** units into only four different rectangles. The *factors* of 24 are 1, 2, 4, 6, 8, 12 and 24.
24 is also a **composite number**.

We can only arrange **7** units into one rectangle!
The *factors* of 7 are 1 and 7.
7 has only two factors so it is a **prime number**.

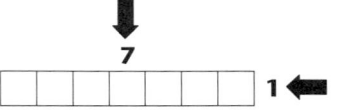

Note: We can only make one rectangle with any prime number.

Use the grid below to find the factors of 18. List the factors in the box below. Then use graph paper to find the factors for 5, 21, 13, 36, 40, 27, 19.

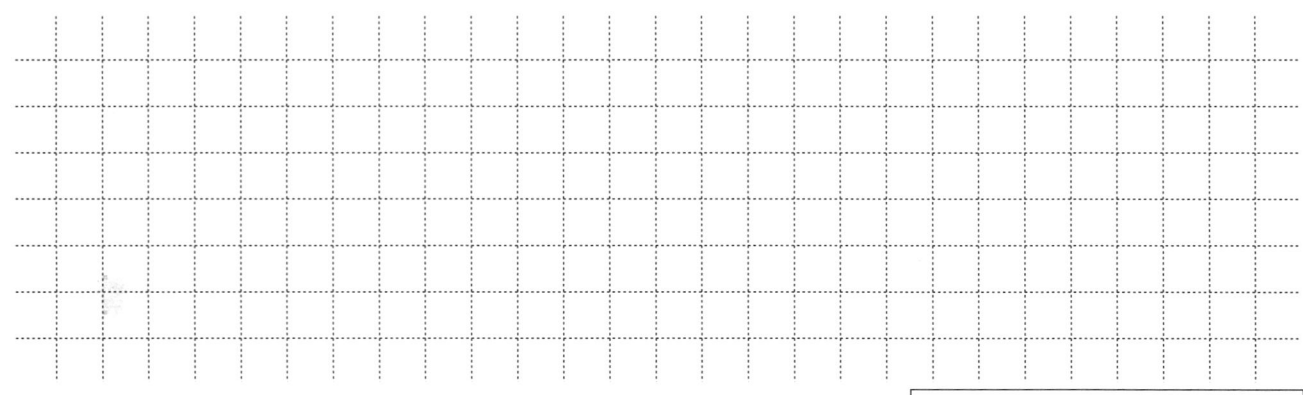

Factor Patterns

A factor is a whole number (or integer) which divides into another whole number exactly. For example: 3 is a factor of 12
5 is a factor of 25

The 8 x 8 grid on the right can be used to find factors of 8.
First, we shade all the squares down the first column (multiples of 1).
Then we shade multiples of 2 (2x table) in the second column. Then we shade multiples of 3 (3x table) down the third column and so on to the eighth column.

When we have done all eight columns, the bottom line shaded squares (shown by an arrow ←) give us the factors of eight.

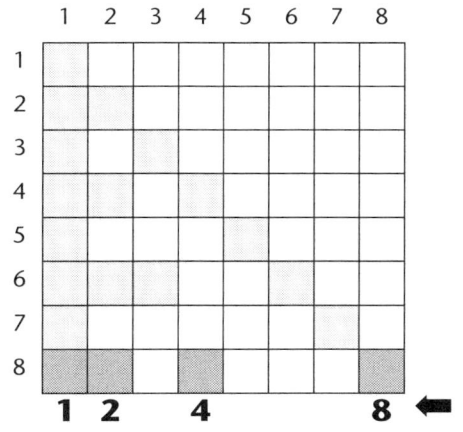

Factors of Eight

Now It's Your Turn!
Using the 12 x 12 square below, find the factors of 12.

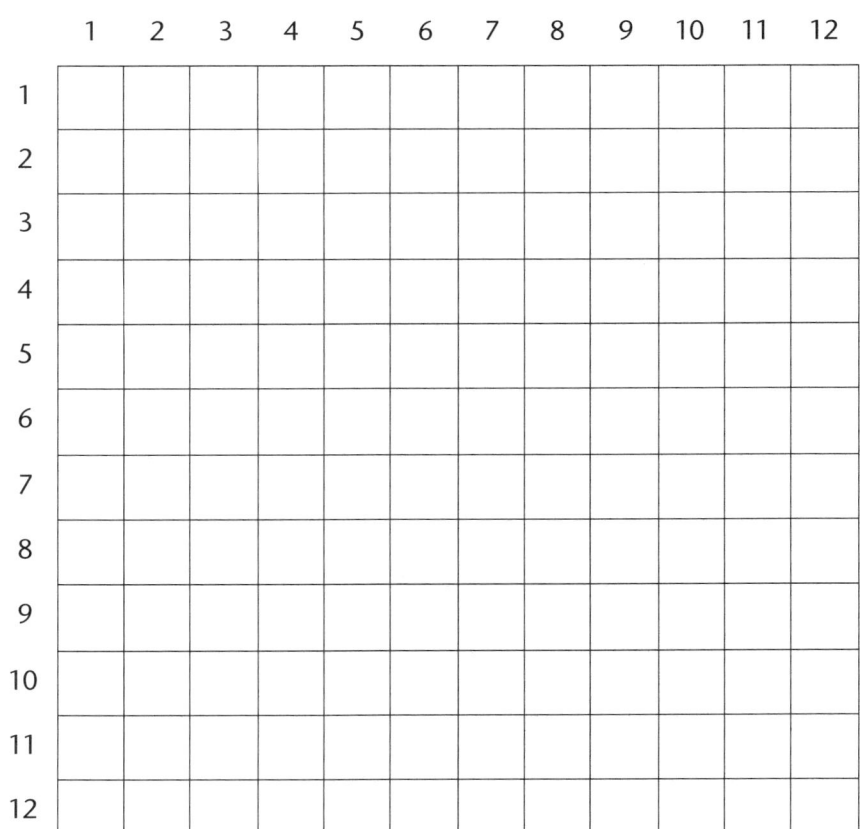

Now make up a 20 x 20 grid with 1-cm squared paper and use it to find the factors of 20.

Products and Isometric Shapes N

When we multiply two or more numbers together, for example, 3 x 5 or 2 x 5 x 7, the answer is called a **product**. Find the products of the numbers below and then shade the answers in the boxes. Your shaded answers will form a shape. Pick the name of this shape from the list at the bottom of the sheet and write it in the box below the shape. Check that the shape fits the clues.

Shape A

8 x 20
3 x 3
2 x 3 x 2
1 x 1 x 1
5 x 9
3 x 18
7 x 13
9 x 7
17 x 0
2 x 5 x 3
5 x 15
6 x 7

29	37	3	20	23	6	80	5	66
2	23	7	4	5	37	90	65	7
39	5	37	23	17	11	80	2	41
11	8	13	160	30	45	8	55	2
24	15	12	11	5	80	63	10	43
13	91	75	54	0	42	9	1	19
8	20	10	70	17	29	37	23	17
31	57	2	40	33	60	29	77	16
7	13	35	31	40	19	52	8	23

Clues
1. Two parallel sides.
2. Two angles > 90°.
3. Two angles < 90°.

Shape A is a:

Shape C

7 x 8
12 x 7
14 x 7
8 x $\frac{1}{2}$
6 x 16
9 x 6
2 x 7 x 3
8 x 1 x 0 x 2
9 x 9
13 x 3
12 x 6
2 x 30 x $\frac{1}{2}$
14 x 4
6 x 15

17	37	40	25	34	19	7	15	34
10	50	1	22	35	81	54	72	42
25	11	17	10	0	60	40	98	10
60	7	15	4	35	37	56	17	19
19	35	30	34	7	39	50	29	40
22	96	90	84	56	22	6	1	53
11	15	40	25	29	37	91	7	60
50	34	50	1	53	35	17	29	22
53	7	25	60	17	10	11	34	15

Clues
1. Four vertices.
2. Opposite sides parallel.
3. No right angles.

Shape C is a:

Shape B

6 x $\frac{1}{2}$
2 x 3 x 0 x 4
5 x 18
7 x 7
6 x 13
5 x 14
1 x 1 x 1 x 1
11 x 5

9	28	50	15	20	8	23	45	10
7	50	38	37	15	64	38	26	17
19	32	6	90	55	49	55	14	20
17	19	5	78	29	70	11	23	60
29	31	13	1	0	3	23	80	6
12	8	7	2	51	34	19	80	62
53	18	13	14	40	4	2	81	11
2	31	19	6	19	6	16	5	9
27	40	17	4	13	26	12	23	8

Clues
1. Four congruent sides.
2. A special rectangle.
3. Four right angles.

Shape B is a:

Shape D

7 x 13
1 x 4 x 1 x 1
40 x $\frac{1}{2}$
3 x 23
6 x 11
8 x 5
3 x 0 x 2 x 5
17 x 5
1 x 71 x 1
5 x 2
1 x 1 x 5
94 x $\frac{1}{2}$

44	59	14	9	11	32	3	43	88
40	29	1	79	48	43	13	7	1
38	1	40	5	91	45	18	88	38
31	11	66	57	77	47	65	2	21
67	72	69	46	60	1	4	52	37
13	41	85	86	7	10	84	34	53
73	31	71	0	20	2	97	17	46
1	44	4	38	99	60	11	37	4
23	23	8	11	14	77	13	29	8

Clues
1. Five sides.
2. Five vertices.
3. Two right angles.

Shape D is a:

Shapes: Pentagon Oval Square Trapezium Triangle Parallelogram Rhombus

Odd Numbers

1. Describe what an odd number is.

   _____

   _____

2. Circle the odd numbers in the numbers below.

       17   72   84   89   276   375   841   968   1253

   Explain how you knew which numbers were odd.

   _____

   _____

When you multiply two odd numbers the answer, or *product*, is always odd.

3. Test the statement above by completing the product grid on the right. Rewrite the statement in the box below if you find it's true.

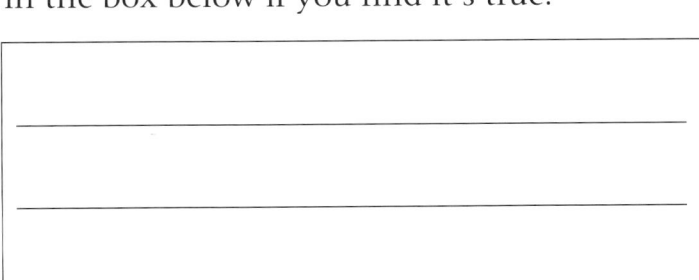

4. Does this work when we multiply three odd numbers? Find the products below.

   (a) 3 x 1 x 5 =   15   (Answer is odd)

   (b) 7 x 3 x 3 = _____ (_____)

   (c) 9 x 5 x 11 = _____ (_____)

   (d) 3 x 5 x 7 = _____ (_____)

   (e) 1 x 7 x 5 = _____ (_____)

   (f) 3 x 3 x 3 = _____ (_____)

   (g) 5 x 5 x 5 = _____ (_____)

   (h) 3 x 1 x 15 = _____ (_____)

Prime Numbers

A prime number is a number which has only two *factors* (numbers which divide into it exactly).
The factors of 7 are 1 and 7. No other numbers divide into 7 exactly so 7 is a prime number.
Now shade in all the prime numbers in each set below and you will spell a math term you should know in each line!

Remember! One is not a prime - it only has one factor.

A.

2	7	13	11	3
17	6	1	4	23
2	3	2	11	19
7	9	10	12	8
11	10	20	15	10

7	13	29	7	3
3	8	12	9	17
5	3	2	7	23
13	16	14	5	20
17	8	1	10	11

8	6	7	1	12
20	8	31	10	15
40	10	11	18	21
1	9	13	6	9
4	24	23	30	10

2	17	11	7	2
5	10	6	8	1
3	7	2	13	23
20	30	8	10	7
29	19	3	2	5

7	6	8	4	17
11	2	10	7	19
3	4	5	15	3
2	10	4	20	5
13	15	8	9	7

B.

8	10	7	20	14
6	3	8	2	1
2	6	12	6	11
7	43	13	17	3
3	1	10	8	2

5	13	2	7	11
2	10	6	8	19
7	23	29	5	3
11	12	15	7	10
19	20	18	6	2

7	2	3	11	17
19	10	12	40	3
7	23	17	11	29
5	21	20	7	12
2	8	6	10	3

10	8	11	12	6
4	7	1	2	10
3	15	20	6	5
11	19	7	3	13
17	8	10	1	2

7	6	10	8	5
1	11	4	2	6
6	14	17	20	30
8	10	11	16	9
4	6	3	10	8

C.

7	5	2	11	3
10	1	5	6	8
12	4	17	9	15
8	6	13	12	20
4	1	11	10	8

10	6	11	4	1
6	13	20	17	12
7	8	9	15	3
13	11	19	7	2
5	12	16	9	5

2	4	8	10	12
7	20	1	6	10
11	15	20	9	12
13	18	21	24	15
5	2	7	3	17

3	10	1	6	8
11	12	14	16	20
17	1	6	9	10
23	21	40	8	15
5	3	2	7	11

7	1	6	8	5
10	17	20	13	10
14	16	29	25	30
8	10	3	4	12
6	1	2	9	14

D.

11	7	2	5	3
2	10	6	8	13
17	3	2	7	11
23	1	12	11	14
7	18	20	30	3

8	10	23	4	6
12	17	15	19	21
23	28	30	4	37
29	11	13	17	5
41	8	6	15	2

2	3	7	8	10
31	6	8	5	12
23	8	12	15	11
7	20	25	17	8
5	11	13	30	6

10	6	2	50	26
20	8	7	24	40
12	30	11	14	22
15	20	23	8	6
16	12	31	10	9

8	10	37	1	16
12	6	11	4	25
9	15	7	40	30
6	1	5	24	8
4	18	2	21	10

World Teachers Press — Problem Solving Through Investigation – Book 1

Composite Numbers

A **composite number** has more than two factors. **Factors** are numbers which divide into another number exactly. For example: the factors of the composite number 12 are 1, 2, 3, 4, 6 and 12.

Shade the composite numbers in the grid below to discover two solid shapes. Then complete the sentences. Do not shade prime numbers as they have only two factors.

5	1	7	1	3	7	2	5	1	11	7	13	7	2	17	7	3	11	13	2	13	11	17	2	13	11
2	7	11	11	19	13	11	2	2	7	1	11	17	23	1	5	11	19	1	3	7	1	17	1	5	2
7	**A**		7	5	1	13	11	2	3	13	5	3	19	11	3	1	7	13	2	5	7	2	19	7	7
11			5	7	2	3	5	2	15	6	20	8	10	7	2	2	11	2	3	7	5	5	7	2	11
1	5	2	5	11	2	5	11	9	3	11	7	16	9	11	5	1	5	2	1	2	10	3	5	17	7
11	7	3	1	7	11	5	6	2	1	1	15	5	6	7	3	7	11	13	11	9	1	30	5	2	3
5	11	2	7	3	17	4	11	19	13	6	23	5	4	3	2	19	3	13	4	11	13	7	18	5	2
1	19	3	11	23	10	7	3	5	10	2	7	18	11	13	7	13	5	6	2	2	5	13	17	6	11
17	7	11	19	8	2	1	11	12	17	5	20	17	23	7	5	17	10	2	3	7	13	29	4	5	7
11	1	3	4	19	2	19	9	3	2	25	19	23	2	2	3	30	7	13	1	5	1	16	7	3	1
7	11	6	14	10	50	4	3	7	8	17	1	7	11	17	8	17	5	3	7	13	14	2	3	5	7
19	2	8	7	2	13	24	5	10	7	23	3	7	13	6	13	2	1	17	3	10	11	7	1	2	13
1	11	12	3	11	5	21	6	2	1	3	5	11	10	3	20	17	3	11	12	5	3	11	5	11	1
11	13	10	16	8	18	15	23	7	3	17	13	9	2	13	3	12	7	6	5	11	2	**B**		7	5
7	2	7	7	1	13	5	29	1	2	17	15	12	8	20	18	6	8	5	13	1	7			11	7
7	11	1	11	5	7	2	7	5	7	11	13	1	5	2	1	17	2	1	17	11	2	5	1	7	3

Remember, since number 1 has only one factor it is neither prime nor composite.

Shape **A** is a _____ and has _____ faces, _____ vertices and _____ edges.

Shape **B** is a _____ and has _____ faces, _____ vertices and _____ edges.

World Teachers Press — Problem Solving Through Investigation – Book 1

Prime and Composite Numbers N

A mathematician called Goldbach stated that *every* even number greater than 2 can be found by adding together two prime numbers.

For example: 10 (even number) = 7 (prime number) + 3 (prime number)
                8 (even number) = 3 (prime number) + 5 (prime number)

1. Explain what an even number is.

   _____

   _____

2. Explain what a prime number is.

   _____

   _____

3. Now take the even numbers below and discover if Goldbach's statement works for them. Write your answers as shown in the examples above. There may be more than one answer for each number, but just give one.

   Remember: The number 1 is not a prime number as it has only a single factor.

   26 (even number) = _____ (prime number) + _____ (prime number)

   32 (_____) = _____ (_____) + _____ (_____)

   44 (_____) = _____ (_____) + _____ (_____)

   56 (_____) = _____ (_____) + _____ (_____)

   80 (_____) = _____ (_____) + _____ (_____)

   22 (_____) = _____ (_____) + _____ (_____)

   20 (_____) = _____ (_____) + _____ (_____)

   48 (_____) = _____ (_____) + _____ (_____)

   16 (_____) = _____ (_____) + _____ (_____)

   102 (_____) = _____ (_____) + _____ (_____)

   64 (_____) = _____ (_____) + _____ (_____)

No one has ever found that this statement is wrong – if you do, you will be famous!

Commutative Law

Math is often concerned with patterns in numbers. Addition and multiplication are both **commutative** because 2 + 7 = 7 + 2 and 3 x 10 = 10 x 3.

The two arrays below show this as each array has a symmetrical number pattern on either side of the leading diagonal.

+	1	2	3	4
1	2	3	4	5
2	3	4	5	6
3	4	5	6	7
4	5	6	7	8

x	1	2	3	4
1	1	2	3	4
2	2	4	6	8
3	3	6	9	12
4	4	8	12	16

Because addition and multiplication are commutative, it doesn't matter whether you say 4 + 2 or 2 + 4, 2 x 4 or 4 x 2 along the rows in the arrays, you will still arrive at the same answer.

1. Complete these addition and multiplication arrays and draw in the diagonal. Then shade to show the patterns in the numbers. Shade the same numbers in the same color.

+	3	4	5	6
3				
4				
5				
6				

x	3	4	5	6
3				
4				
5				
6				

2. Now make up your own arrays below. If you pick large numbers, use a calculator to save time.

+				

x				

Square and Triangular Numbers

A **square number** is a number whose units can be arranged into a square.

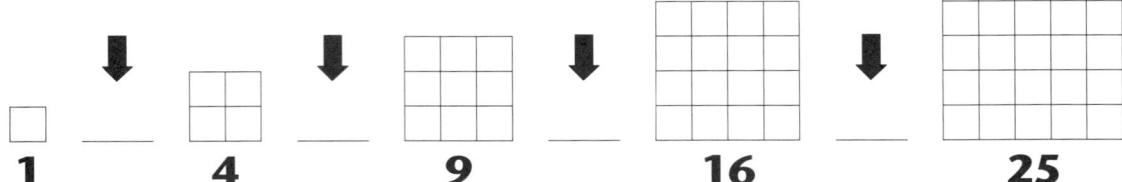

A **triangular number** is a number whose units can be arranged into a triangle.

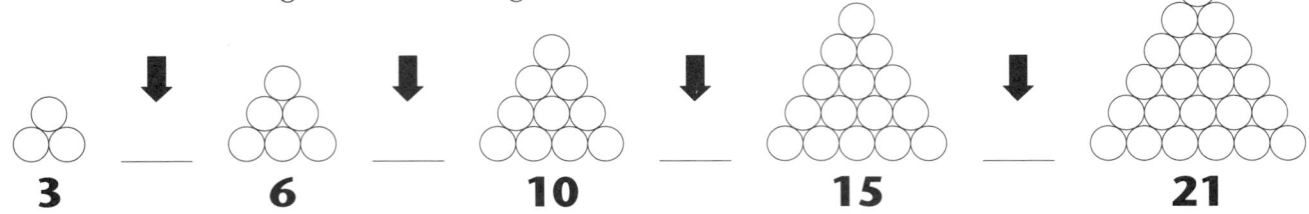

1. In the blanks above, write down the difference between adjacent numbers to **find a pattern**.

This is an **array** of the **square number** 9.

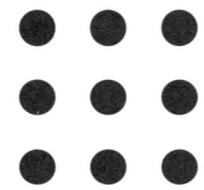

Here is the same array showing that the square number 9 is made up of two adjacent triangular numbers.

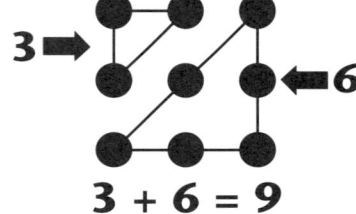

2. Do the same to the square number arrays below.

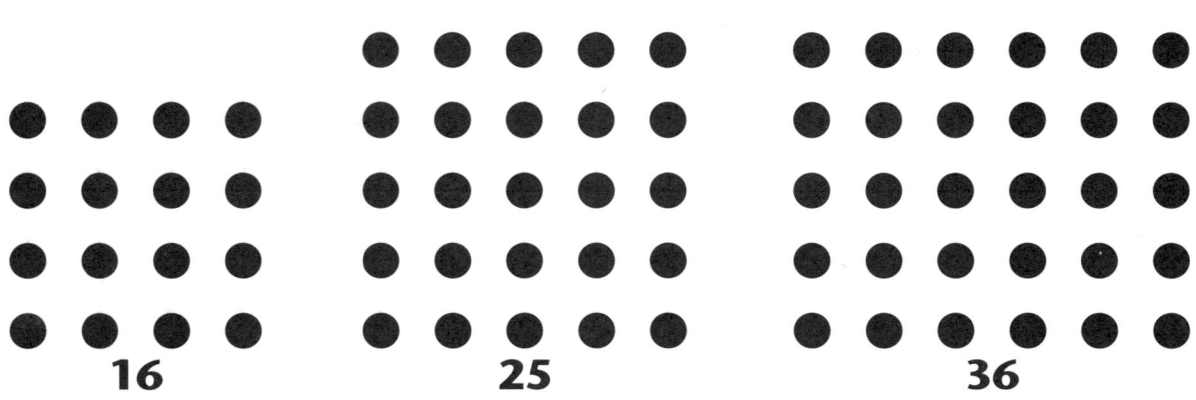

3. Now pick your own square number and do the same on a piece of graph paper.

Probability

If you tossed a coin once it could come up **heads** or **tails**. This means there is one chance in two of tossing a head and one chance in two of tossing a tail. If you tossed a coin 100 times, how many heads and tails do you think you would toss?

Write down your estimate:

I think I would toss _____ heads and _____ tails.

Now you and your partner toss a coin 100 times and record your results on the hundred grid to the right. When the coin comes up as a head, write an **H** in the square. When the coin is a tail, write a **T** in the square.

Example: | H | T | T | H | T | H | H |

When you have filled the grid, count up the Hs and write the number of heads tossed in the space provided below the grid. Do the same for the tails.

Are your results **exactly** the same as your estimate? _____

Are your results **approximately** the same as your estimate? _____

Total Heads: _____   Total Tails: _____

Now repeat your 100 tosses on the second hundred grid to the left.

Are your results approximately the same? _____

Do you think you could predict the exact number of heads and tails? _____

Why or why not? _____

_____

_____

Total Heads: _____   Total Tails: _____

World Teachers Press — Problem Solving Through Investigation – Book 1

Magic Squares

The idea of a magic square appears to have come from China and one has been found in a document dated hundreds of years before the birth of Jesus Christ.

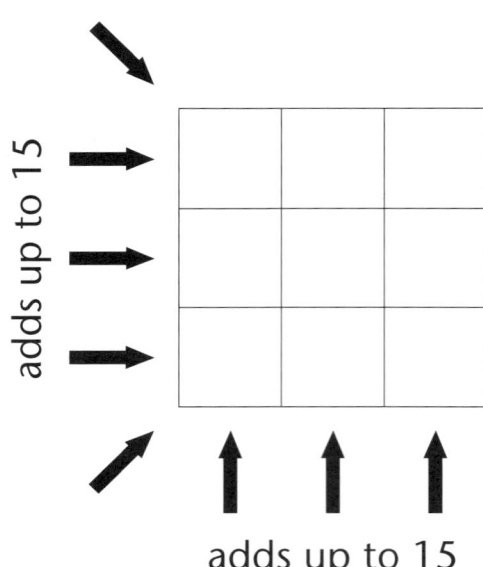

2. In the square above use **only** the numbers **4, 5** and **6** to complete the square. They must add up to the number 15 along the rows, columns and diagonals. Each number must be used **exactly three times**.

1. Use the numbers **1, 2, 3, 4, 5, 6, 7, 8** and **9** in the square below so they add up to the number 15 in all directions (indicated by arrows).
   Use each number only once.
   A legendary Chinese emperor is said to have seen this square carved in the shell of a sacred tortoise.

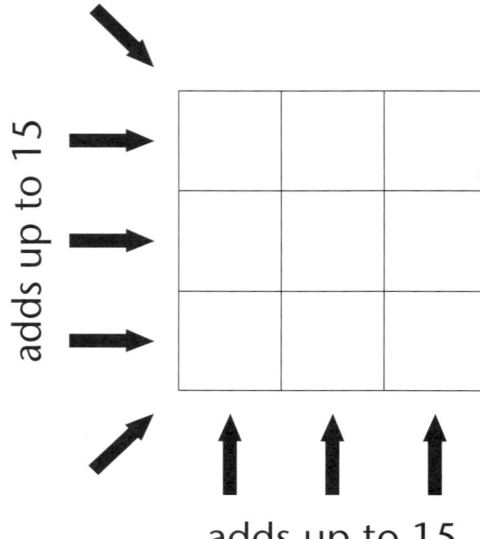

3. The first magic square devised by a European was done by German artist Albrecht Durer in the 16th century. He used the numbers **1 to 16** inclusively to make a magic square that added up to **34** in all directions. Complete Durer's magic square on the left, using each number, from 1 to 16, **only once**. Some numbers are already given to help you. When you have completed this magic square, answer the following questions.

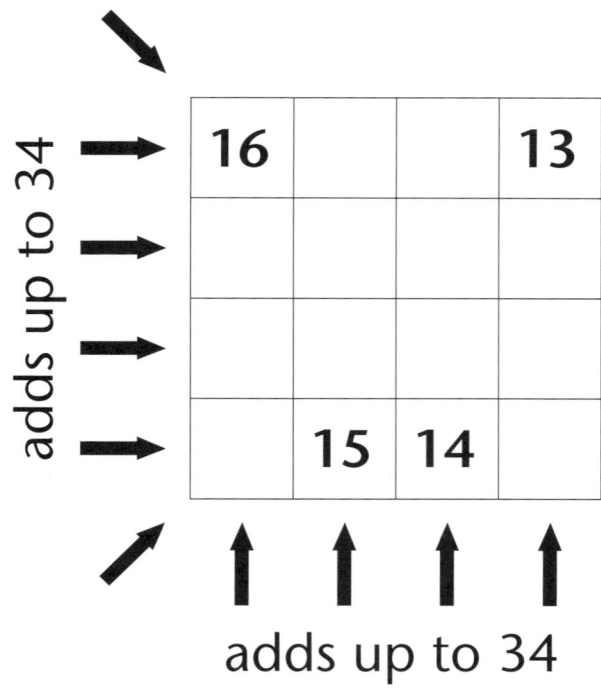

(a) What is the sum of the four corner numbers?_____

(b) What is the sum of the four numbers in each quarter of the square?

_____, _____, _____, _____

Fun with the Magical 9!

1. Divide each of the following numbers by **9** and then write the **remainder** under the box.

   | 16 | 78 | 371 | 672 | 2,183 | 13,351 |

   Now find the **sum** of the numerals of each boxed number as **a single number**.
   For example: 257 = 2 + 5 + 7 = 14 = 1 + 4 = 5. Put this single numeral above its box.
   Now write about what you discovered.

   _____

   _____

   This magic only works if there is a remainder!
   Try your own number using eight digits.

Use your calculator for the next two pieces of magic!

2. Each number below has four **different** digits. Rewrite each number using the same digits but make the new number smaller than the original one.
   For example: 5,328 could become 2,358, 3852, 3528, etc.

   Subtract the new number from the old number and then divide your answer by **9**.

   8,253   9,246   7,345   6,287   5,471

   9)   9)   9)   9)   9)

   Write about what you discovered.

   _____

   _____

3. Now take the three-digit numbers below and reverse the digits. For example: 752 becomes 257. Subtract the new number from the old number and divide the answer by 9.

   835   926   741   843   612

   9)   9)   9)   9)   9)

   Write about what you discovered.

   _____

   _____

Operations – Calculator Work

This activity will give you practice on your calculator. Find the sums, then turn the calculator upside down and read the word in the display window. If there is no word, do your calculation again! Do the calculations in brackets first.

Clue	Calculation	Answer
1. A difficult tennis shot	(8 x 20 x 5) + 7	_____
2. Fish breathe with this	(5,787 ÷ 3) x 4	_____
3. A best selling book	191 x 11 x the product of 2 and 9	_____
4. Used in cold countries	485,375 - 24,000	_____
5. A ship's records are kept in it	(707 - one century)	_____
6. A fish or part of a shoe	3 x 5 x 247	_____
7. You'd be upset by this	(L x CX) + VII	_____
8. It has to be repaired sometimes	300 tens + XLV	_____
9. Used in the garden	35 hundreds and 4	_____
10. A bird has one	(40 x 200) - 200 - 82	_____
11. To walk as if injured	400 thousand -21,196	_____
12. A marsh or swamp	(12 x L) + VIII	_____
13. Your ear has one	half of 7614	_____
14. Shouldn't be told	15 score + XVII	_____
15. You'll find them in a gaggle	(35 x 1,000) + 336	_____

Note:
Never accept that the calculator answer is always correct. You may have pressed the wrong button! Always estimate the answer first and then compare your estimate with the calculator answer.

Number Patterns (1)

In the first circle **multiples of three** (3 x table) are arranged around the circumference. Start at 0 and then rule a line to 3, then from 3 to 6, then from 6 to 9, and so on. You will end up with a pattern. Color all of the **triangles** and **quadrilaterals** in this pattern. Then do the same with the other circles. Be sure that **adjacent shapes with a common side** are not the same color!

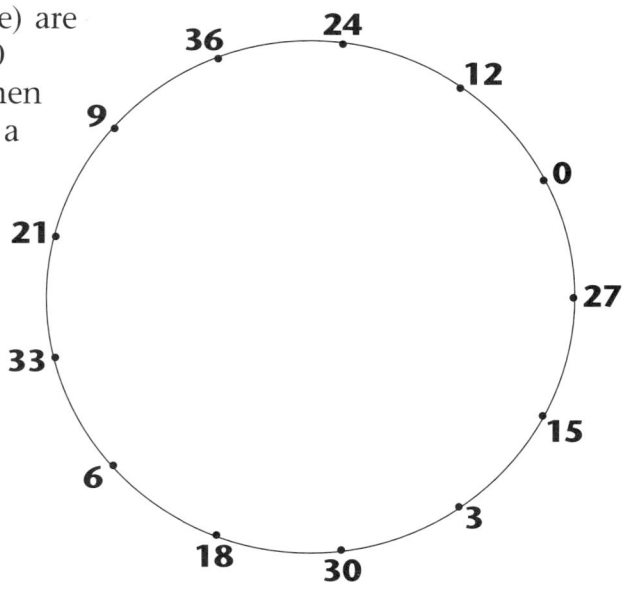

Multiples of three

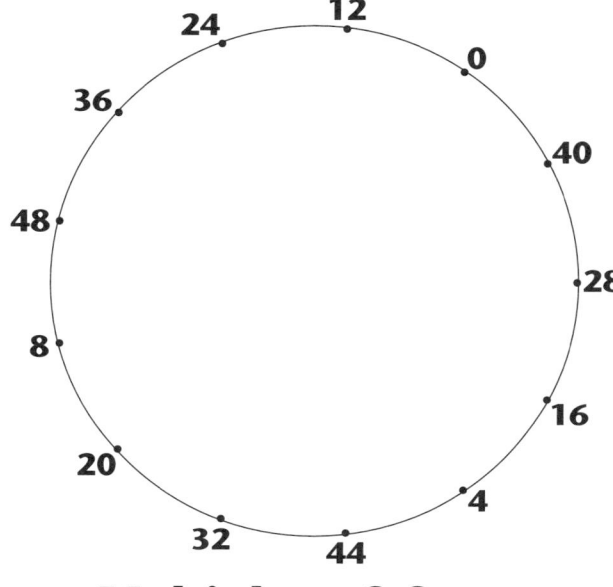

Multiples of four

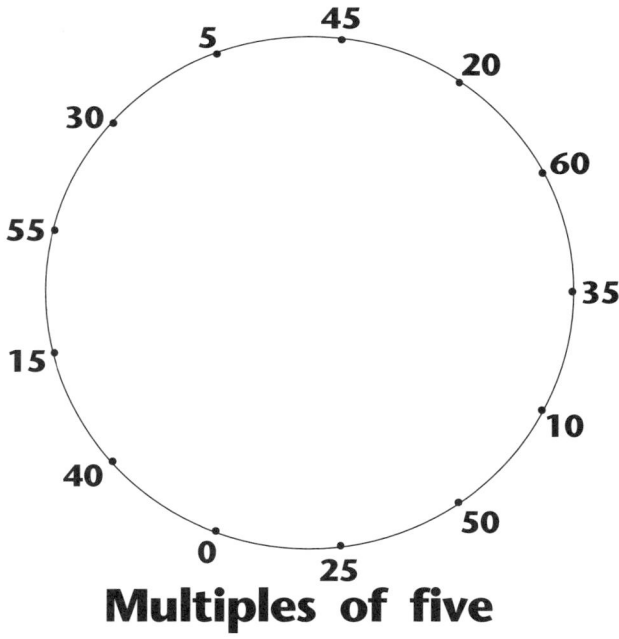

Multiples of five

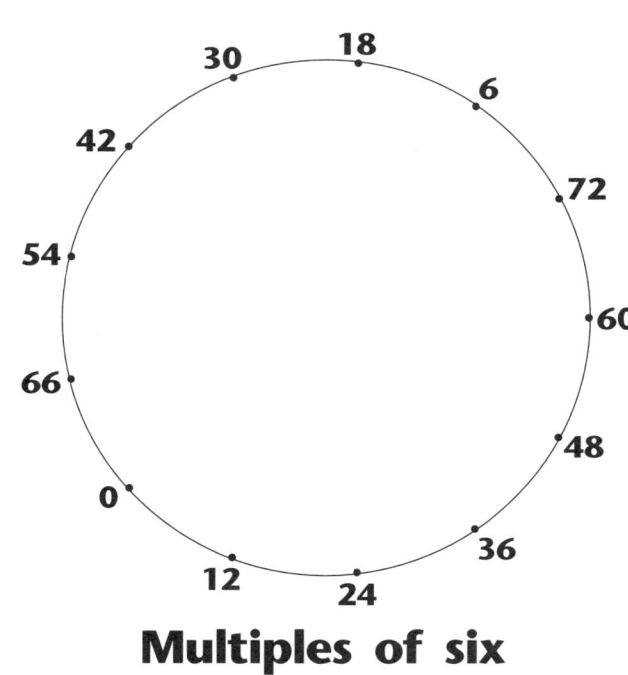

Multiples of six

Look at each circle and write down the differences between adjacent numbers. Write a comment about these secondary patterns.

_____

_____

Remember, start at 0, finish at 0!

Number Patterns (2)

N

Triangular numbers are numbers whose units can be arranged into triangles.

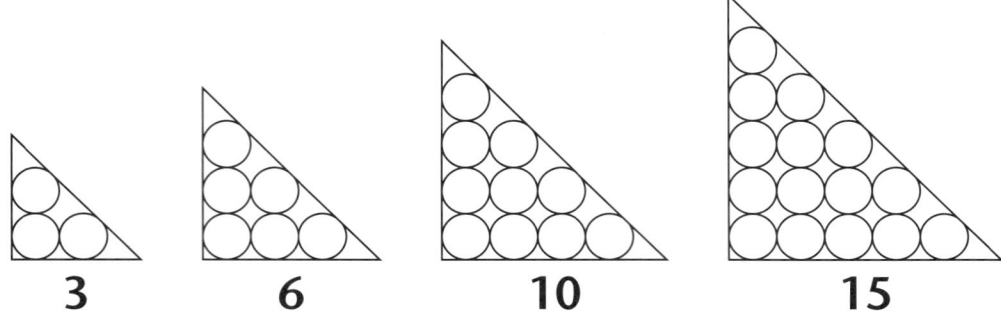

3   6   10   15

Can you see a pattern?

Diophantus, a Greek mathematician, was interested in patterns in numbers and discovered one pattern described below.

Use your calculator and multiply each triangular number in the circles by 8 and then add 1. The first has been done for you.

(28)	(36)	(45)	(55)	(66)	(78)	(91)	(105)
x 8	x 8	x 8	x 8	x 8	x 8	x 8	x 8
= 224	=	=	=	=	=	=	=
+ 1	+ 1	+ 1	+ 1	+ 1	+ 1	+ 1	+ 1
= 225	= ☐	= ☐	= ☐	= ☐	= ☐	= ☐	= ☐

Now write what you have discovered about the numbers in the boxes.

_____
_____
_____
_____

Did you find out what Diophantus discovered more than 1,500 years ago?

Number Patterns (3)

Number patterns are found in all areas of math (tables, calendars, square numbers and so on). Use the 100-squares below to find the next three numbers in each series. First of all, color in the squares of the given numbers. Then follow the pattern to color in the next three numbers in the series. Write these numbers in the answer boxes. **Use different colors for each series.**

Example:

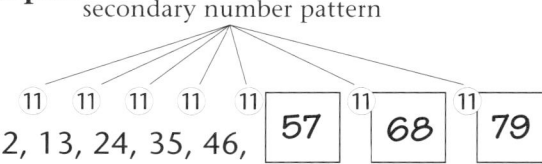

2, 13, 24, 35, 46, 57, 68, 79

Now try these!

1. 72, 83, 74, 85, 76, ☐ ☐ ☐

2. 88, 77, 68, 57, 48, ☐ ☐ ☐

3. 20, 19, 29, 28, 38, ☐ ☐ ☐

4. 11, 2, 13, 4, 15, ☐ ☐ ☐

5. 21, 23, 43, 45, 65, ☐ ☐ ☐

6. 94, 84, 75, 65, 56, ☐ ☐ ☐

7. 1, 12, 23, 34, 45, ☐ ☐ ☐

8. 71, 82, 73, 84, 75, ☐ ☐ ☐

9. 11, 2, 3, 14, 5, 6, ☐ ☐ ☐

Discuss the patterns with your teacher. Did you find the **secondary number patterns** (differences between adjacent numbers) to check your answers?

World Teachers Press — Problem Solving Through Investigation – Book 1

Number Systems

Throughout history people have been able to count their sheep, cattle, money, measures of wheat, soldiers in their armies, etc. To record their totals they used symbols to represent numbers. **These symbols are called numerals** and they vary from country to country. To record the number of birds here we would use the Hindu-Arabic numeral 5. The Chinese would use 五. Ancient Romans would use the Roman numeral **V**. Listed below are some of the numerals used by ancient peoples. The Egyptians used familiar things for some of their numerals; for example, ඉ (coiled rope) and 𓆼 (lotus flower).

Ancient Egyptian

1	/
3	///
5	/////
10	∩
100	ඉ
1,000	𓆼
10,000	ᒉ
100,000	ᗢ

The Ancient Egyptians usually put their smaller numbers to the left, so 13 was written as ///∩.

Ancient Roman

1	I
3	III
4	IV
5	V
9	IX
10	X
40	XL
50	L
100	C
500	D
1,000	M

Ancient Babylonians

1	▼
3	▼▼▼
7	▼▼▼▼ / ▼▼▼
10	◄
13	◄▼▼▼
20	◄◄
41	◄◄◄◄ ▼
50	◄◄◄◄◄
62	▼ ▼▼

The numeral for 1 and 60 was the same, but the 60 numeral was moved to the left (place value).

Central American Mayan

1	•
3	•••
5	—
7	•• / —
10	=
12	•• / =
15	≡
23	••• / =
25	• / =
31	• / ≡
35	≡ / =

When the • symbol was high it stood for 20.

Use these lists to find the Hindu-Arabic equivalents of the numbers below:

1. /∩ᗢ = _____
2. ///∩∩ඉ = _____
3. //ඉඉ𓆼ᒉ = _____
4. /////∩∩𓆼 = _____

1. MDC = _____
2. LVII = _____
3. XCIV = _____
4. XXV = _____

1. ▼ ◄▼▼ = _____
2. ◄◄◄ = _____
3. ◄◄◄▼▼▼ = _____
4. ▼ ◄◄◄ = _____

1. ≡ / = = _____
2. • / ••• = _____
3. • / — = _____
4. •••• / = _____

Index Notation – Calculator Work

This activity will give you practice on your calculator. Find the sums, then turn the calculator upside down and read the word in the display window. If there is no word, do your calculation again.
Do the calculations in the brackets first.

Clue	Calculation	Answer
1. Belonging to him	$(5 \times 10^2) + 14^1$	_____
2. There's nothing in it!	$(37 \times 10^2) + 4^1$	_____
3. A dog can do this	22% of 29×10^2	_____
4. A hardworking insect	$(\tfrac{3}{5}$ of D$) + (6^2 + 2)$	_____
5. A natural protective cover	$(77 \times 10^3) + 300 + $ XLV	_____
6. To recede like the tide	$9^3 + $ CL $ + 4^1$	_____
7. Absolute heaven!	$38^3 + 306$	_____
8. The farmer needs it	$(7 \times $ M$) + 10^2 + $ V	_____
9. A garden tool	$14^2 + 107 + 1^{16}$	_____
10. A high sheen	$(5\tfrac{1}{2}\%$ of $10^6) + 76$	_____
11. Needed by industrial nations	5% of 142×10^2	_____
12. A painful swelling!	$(7 \times 10^3) + (3^3 \times 2^2)$	_____
13. To stare at rudely	$\tfrac{1}{2}$ of $30^2 \times 2^3 + 160$	_____
14. Fluid secreted by the liver	$(20\%$ of $18{,}090) + 10^2$	_____
15. We should all have these!	$(5 \times 100^3) + 318{,}804$	_____

You'll need to know your Roman numerals too!

Mapping

Mapping is a way of showing information clearly, by linking information such as numbers or words. Some mapping examples are given below.

The boxed text between the circles (sets) is called the relation sign.

Why do you think it is given this name? _____
_____

Complete the answers in the mappings below. Then do some of your own mappings.

1.

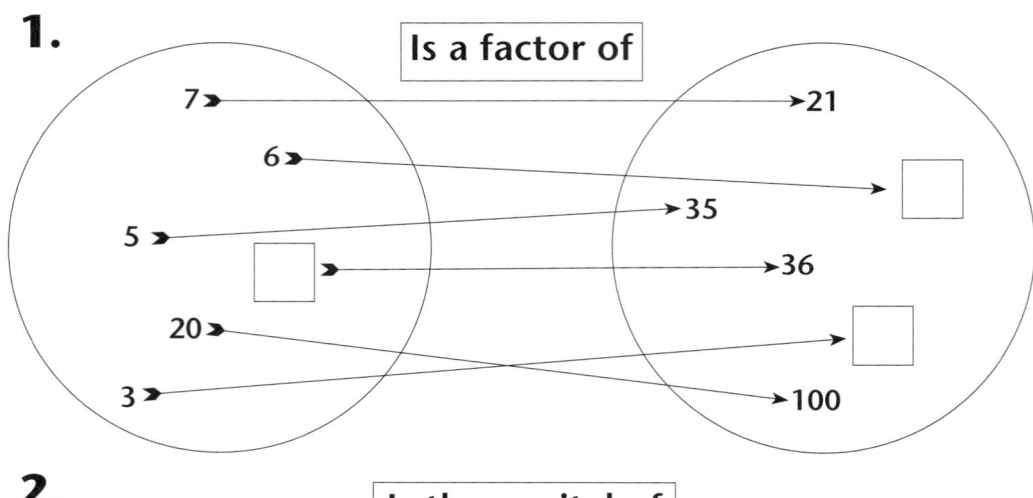

2.

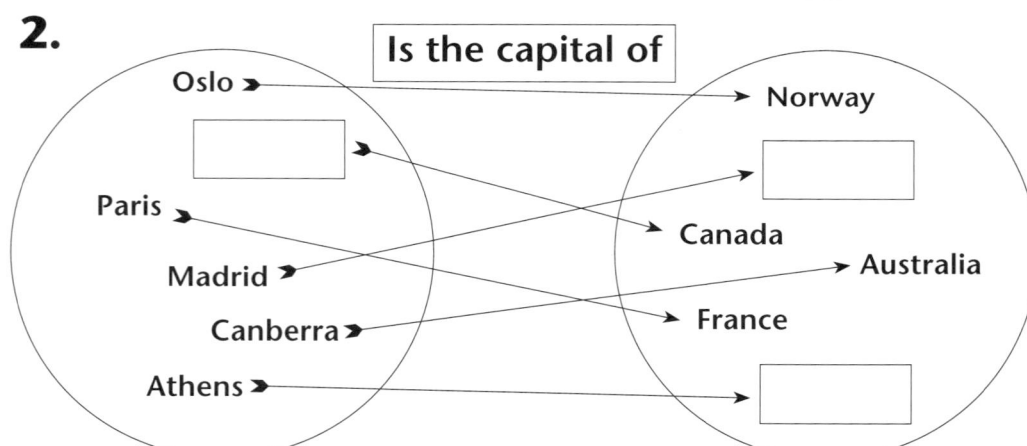

3.

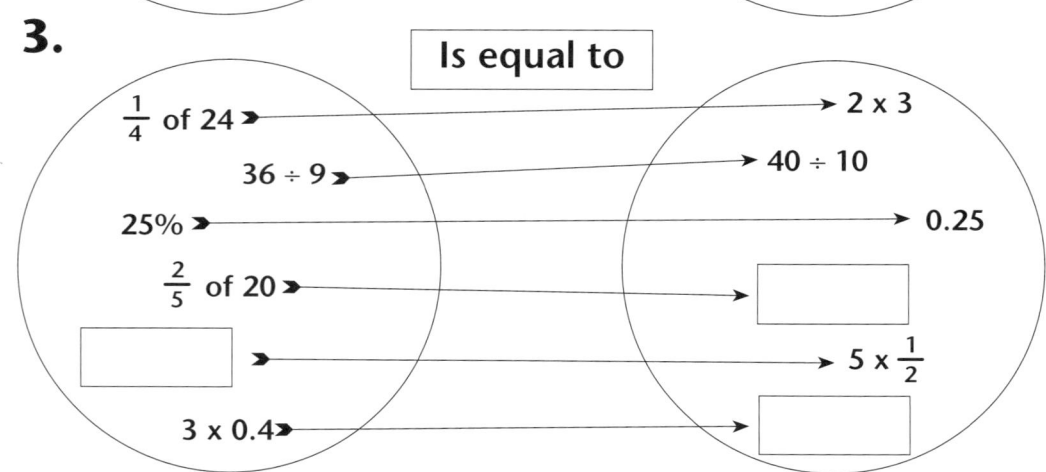

World Teachers Press

Perimeter

The *perimeter* of a shape is the measurement of the distance around the shape. The word comes from the Greek *'peri'* meaning around, and *'metron'* meaning a measure.

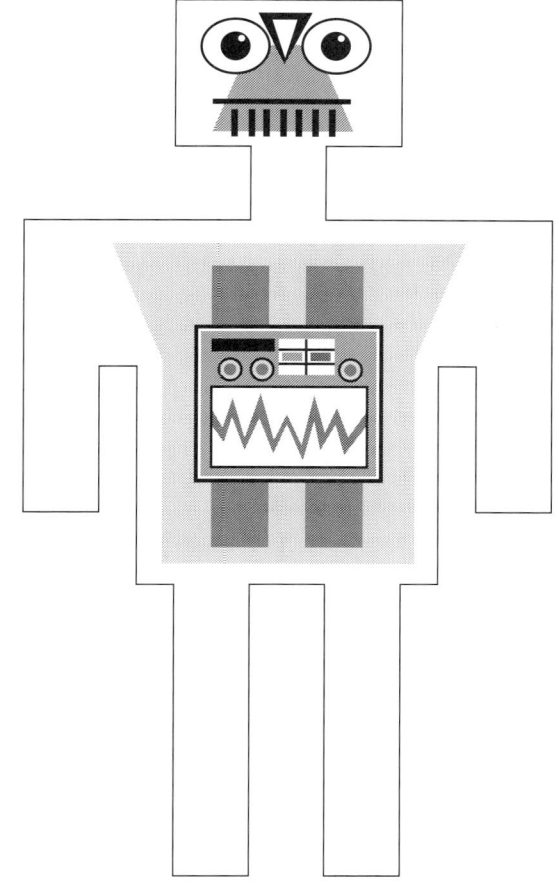

Using a centimeter ruler and a calculator, find the perimeter of this robot with a partner. Then each of you draw your own robot with the same perimeter on 1-cm squared paper.

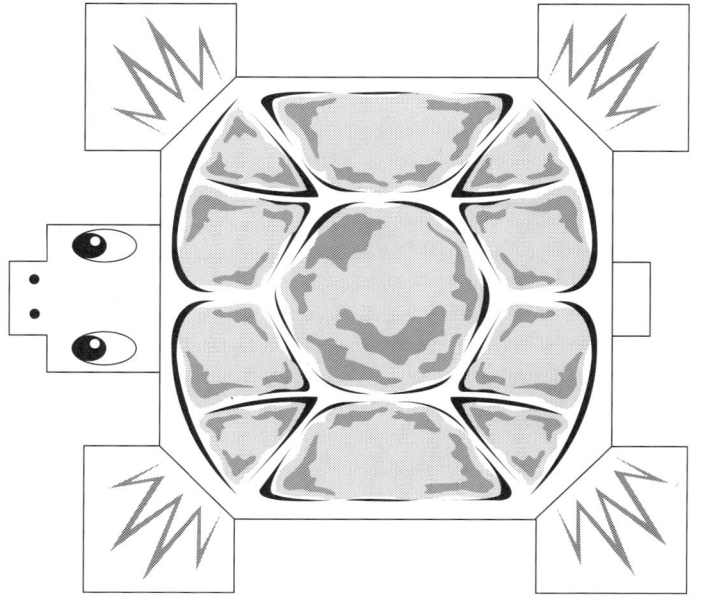

Find the perimeter of this tortoise with your partner as before. Then each of you draw your own animal with the same perimeter.

Color your robot and animal using only **primary** colors.

World Teachers Press — Problem Solving Through Investigation – Book 1

Perimeter and Area   M

Below are five house shapes drawn to scale.

A. 4 m, 14 m

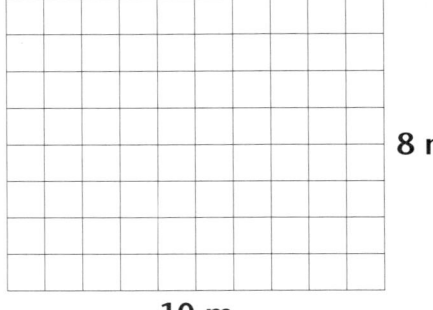
D. 8 m, 10 m

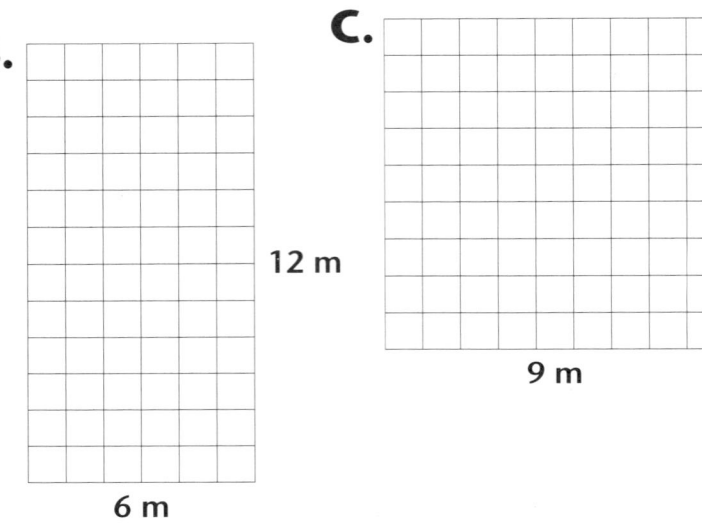
B. 12 m, 6 m
C. 9 m, 9 m

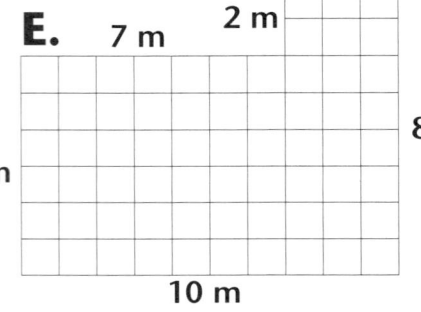
E. 7 m, 3 m, 2 m, 8 m, 6 m, 10 m

1. Find the **perimeters** of the outside walls for each house.

Perimeters

House A = _____ m

House B = _____ m

House C = _____ m

House D = _____ m

House E = _____ m

2. Each little square inside the houses represents a square meter (m^2). Find the **area** for each house.

Areas

House A = _____ m^2

House B = _____ m^2

House C = _____ m^2

House D = _____ m^2

House E = _____ m^2

3. Each home has the same length of outside walls so the bricklayers would be paid about the same amount for each house. Which house would give you the most area for your money spent on the outside walls?

4. Do you think this would be why cheaper brick houses are often square-shaped or close to a square shape? Why or why not?

Triangles

Scalene	Isosceles	Right-angled	Equilateral
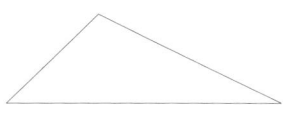	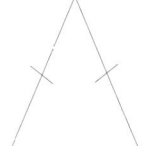	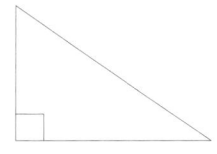	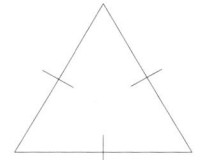
No sides or angles are congruent (equal).	Contains 2 congruent sides and 2 congruent angles.	Contains a right angle (90°).	All sides and angles are congruent.

Look at the pattern of rectangles and triangles below.

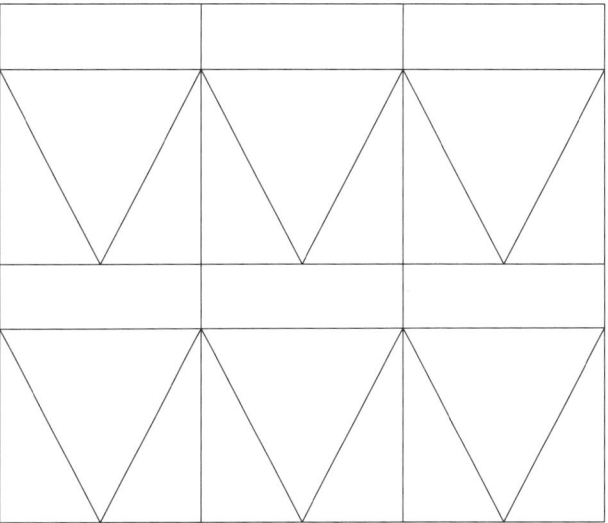

1. Color the small rectangles in the pattern red.
2. Color the right-angled triangles blue.
3. Which kinds of triangles are left?

   _____

4. The base angles of an isosceles triangle

   are _____ .

5. Each angle of an equilateral triangle is

   _____ degrees.

Look at the diagram below.

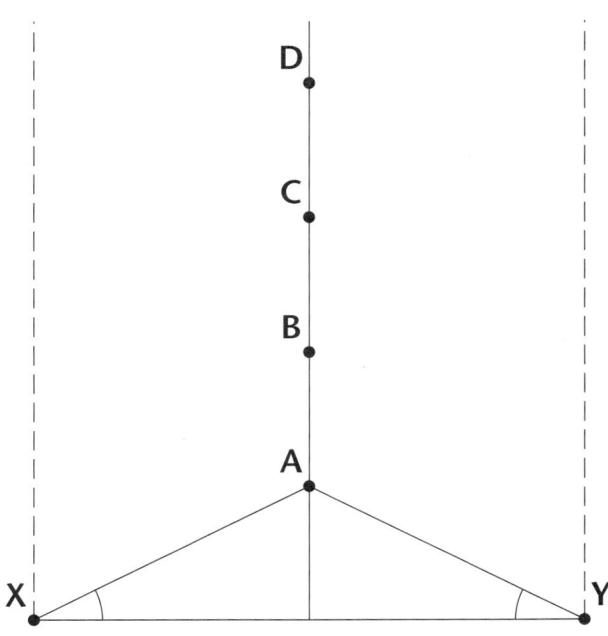

1. Point A has been joined to points X and Y to form an isosceles triangle. Measure the base angles with your protractor. What did you find?

   _____

   _____

2. Now join points B, C, and D to X and Y. Measure the base angles in each new triangle. What did you find?

   _____

   _____

3. Does each triangle you make have two

   congruent sides? _____

World Teachers Press — Problem Solving Through Investigation – Book 1

Angles

M

A. 1. What does **radius** mean? _____

_____

2. What does **diameter** mean? _____

_____

3. What does **circumference** mean? _____

_____

B. 1. Draw a circle with a radius of 4 cm in the box provided below.

Then, as shown in the example diagram below:

2. Draw in the diameter XY on your circle.
3. Pick any two points A and B on the circumference of the circle.
4. Join these two points (A and B) to the end points of the diameter (X and Y).
5. Measure the angles XAY and XBY (marked with a ✭ in the example) using your protractor.

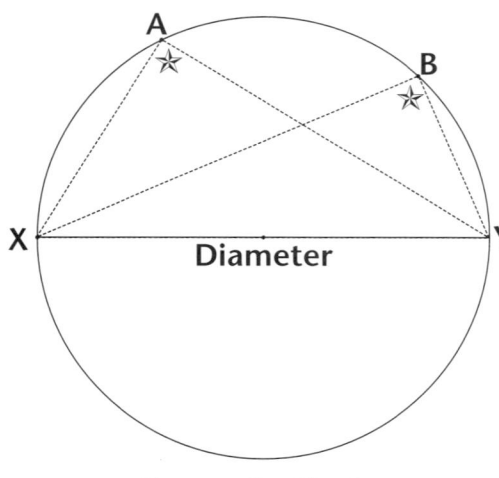

Example Circle

6. What did you find? _____

_____

_____

_____

C. Pick three other points on the circumference of your circle and join each point to X and Y. Measure the three angles. **Are your answers all the same?** _____

World Teachers Press 28 Problem Solving Through Investigation – Book 1

Decades and Centuries

1. How many **decades** are there between Lewis and Clark returning from their expedition to the Pacific Ocean and Alexander Graham Bell demonstrating the first telephone?

2. How many **centuries** are there between the signing of the Magna Carta by King John of England and the defeat of Napoleon at the Battle of Waterloo?

3. How many **decades** are there from Orville Wright's first powered flight in a plane to the first ascent of Mount Everest by New Zealander Sir Edmund Hillary?

4. How many **centuries** are there between the Spanish Armada's attempted invasion of England and the Olympic Games at Seoul?

5. How many **decades** are there between the arrival of the First Fleet in Sydney with convicts from England and the birth of Admiral Richard Byrd (the first man to fly over the North and South Poles)?

6. How many **centuries** are there between the issuing of the Declaration of Independence and the unmanned Viking spacecraft testing Mars for signs of life?

You will need to use an encyclopedia!

Polygons

To prove the three angles of any *triangle* add up to 180°:

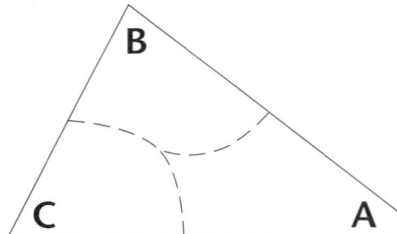

1. Draw any triangle.
2. Label the angles A, B and C and draw in the dotted lines as shown on the left.
3. Cut or tear along the dotted lines.
4. Glue the **vertices** (corners) labeled A, B, C **above the line** ST with each **vertex** (corner) touching point X.

Now do this yourself, and complete the statement below.

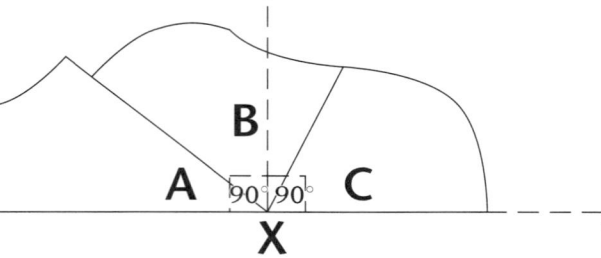

I found that angles A + B + C in my triangle added up to _____°.

To prove the four angles of any *quadrilateral* add up to 360°:

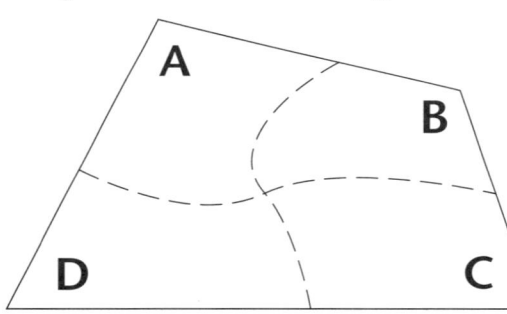

1. Draw any quadrilateral shape. Quadrilaterals have four sides.
2. Label the angles A, B, C and D and draw in the dotted lines as shown on the left.
3. Cut or tear along the dotted lines.
4. Glue the vertices (corners) labeled A, B, C and D so that each vertex (corner) touches point Y.

Now do this yourself, and complete the statement below.

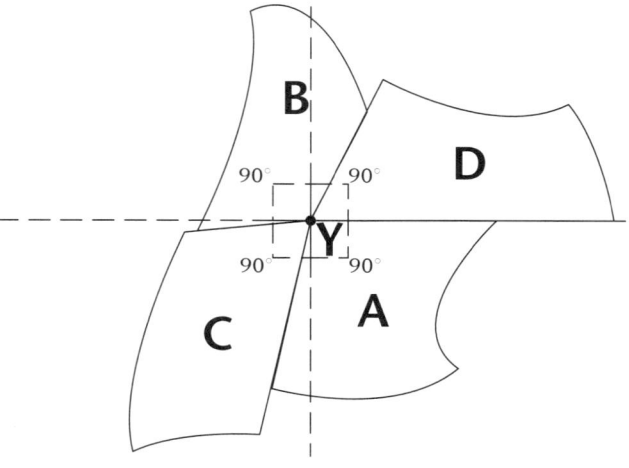

I found that angles A + B + C + D in my quadrilateral added up to _____°.

Duplation

We use the method of multiplication to find the product of two numbers, but this method was not known to the Ancient Egyptians. They used a method called **duplation**, which doubles numbers over and over again.

Look at the example below to see how duplation works.

Problem
A cornfield produces 112 measures of corn. How many measures would be produced by five cornfields?

Duplation (Ancient Egyptian method)

112	from one field
+ 112	from one field
= 224	from two fields
+ 224	from two fields
= 448	from four fields
+ 112	from one field
= 560	from five fields

Multiplication (Our method)

112	from one field
x 5	times five fields
= 560	from five fields

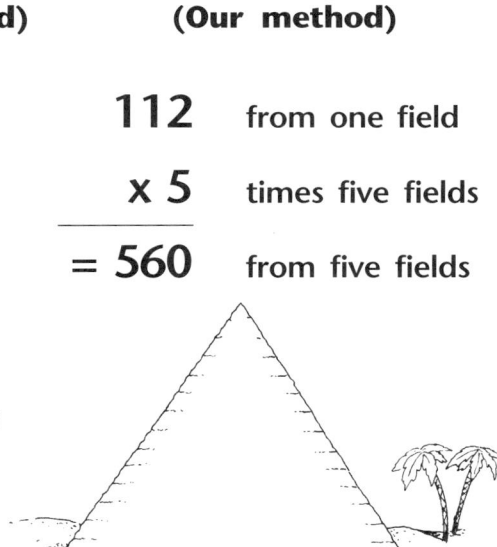

Our method is much quicker!

1. Explain the methods of multiplication and duplation.

    Multiplication: _____

    _____

    Duplation: _____

    _____

2. Work out the following problems using the Egyptian method and check your answers by our method of multiplication.
    Use the problem of the cornfield but instead of 112 measures use:
    (a) 96 measures
    (b) 125 measures
    (c) 150 measures
    (d) 205 measures
    Use the space on the right to do the first problem!

World Teachers Press — Problem Solving Through Investigation – Book 1

Conservation of Area

S

If we have one liter of milk in a plastic container and then pour the milk into a container of a different shape, we still have one liter of milk – its volume is **conserved** (remains the same).

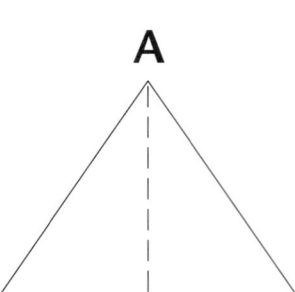
A

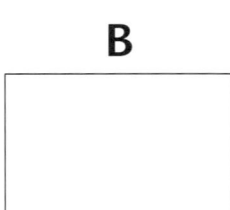
B

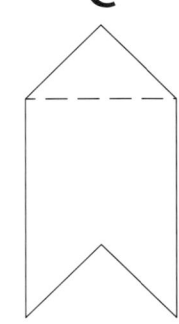
C

We can easily see which is the tallest of the three shapes above, but which shape covers the most area? Cut out shapes **A** and **C** and then **cut along the dotted lines**. Rearrange the two parts in **A** to fit into rectangle **B**. Do the same with the two parts of shape **C**. You should find that they fit exactly. Because they cover equal amounts of paper we say they have **the same area**. Rearranging the parts of one shape to cover the same space as another is an example of **conservation of area**.

Now see if shapes **S** and **T** below fit into the shapes above them.

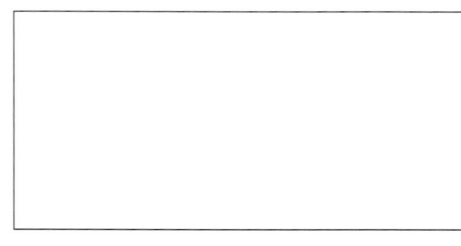

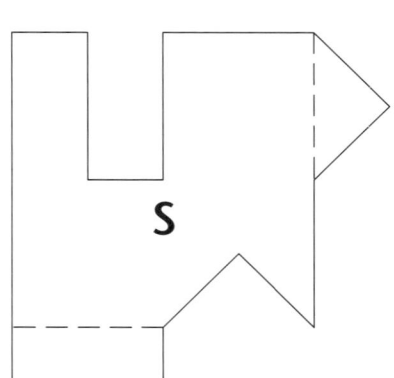
S

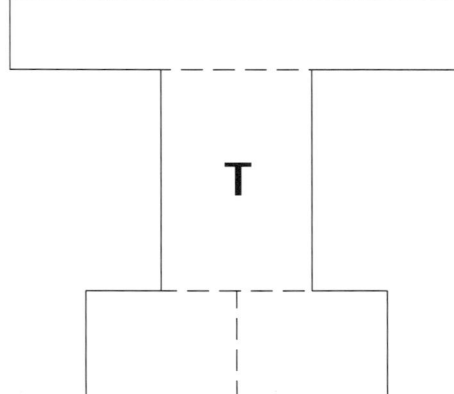
T

World Teachers Press — 32 — Problem Solving Through Investigation – Book 1

Congruency

S

Shapes are **congruent** if they are the same shape and size. Find out if these pairs of shapes are congruent. You may need to measure sides and angles.

Are these shapes congruent?

A. 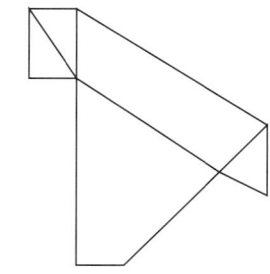 Yes ☐ No ☐

B. 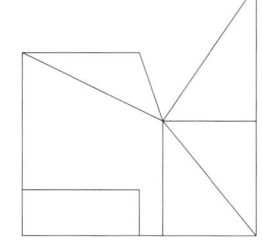 Yes ☐ No ☐

C. 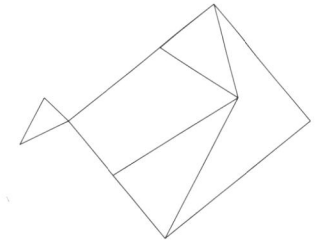 Yes ☐ No ☐

D. 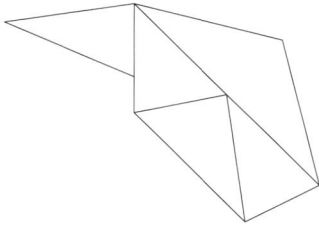 Yes ☐ No ☐

E. 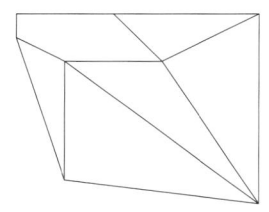 Yes ☐ No ☐

Two of the pairs above are not congruent, but **inside** each pair's main outline there is one pair of congruent, smaller shapes. **Color in the smaller congruent shapes.**

World Teachers Press — Problem Solving Through Investigation – Book 1

Reflective Symmetry   S

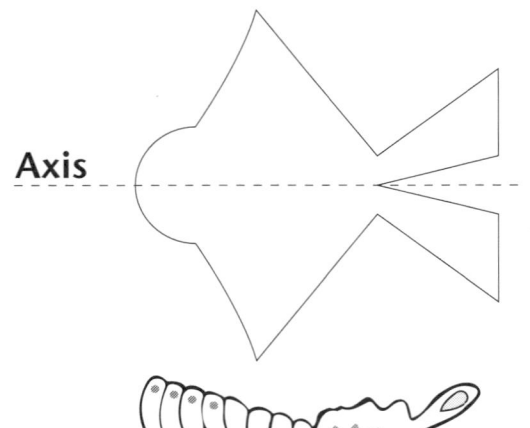

Axis

Look at the two shapes on the left. They are symmetrical because they are the same on either side of an axis of symmetry (or line of balance) drawn down the center. This is known as bilateral (two-sided) or reflective symmetry. A mirror placed along the axis will show an unchanged shape as the image is the same.

Some shapes, squares for example, have more than one line of balance. See the picture below.

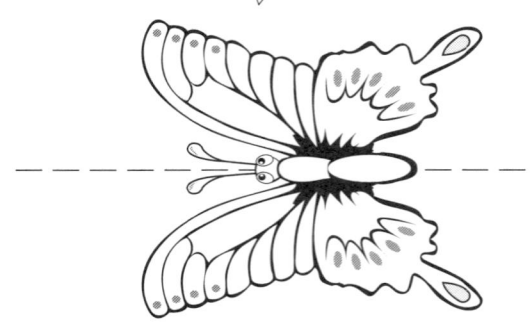

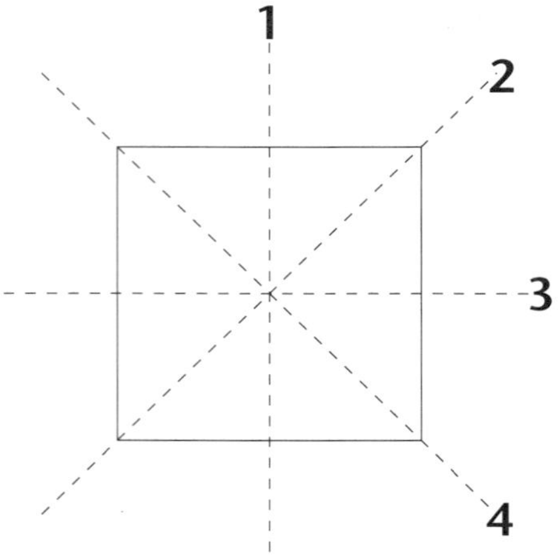

A square has four axes of symmetry.

Fold a piece of paper and draw an outline like the dotted line in the diagram below.

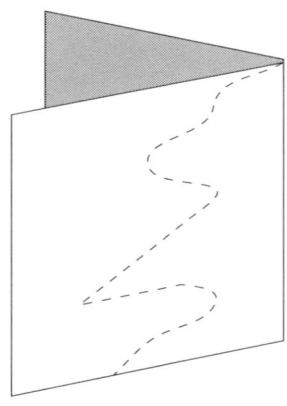

Cut along this line while the paper is still folded and then open up the piece you've cut out. Rule in the axis and shade one half. Try other outlines. You can obtain some interesting shapes with paint blobs inside the fold.

1	2	3

In boxes 1 and 2 draw shapes with more than two axes of symmetry.
In box 3 cut out a symmetrical shape from a magazine. Draw in any lines of balance.

Closed Curves

A **closed curve** has no end-points like these shapes.

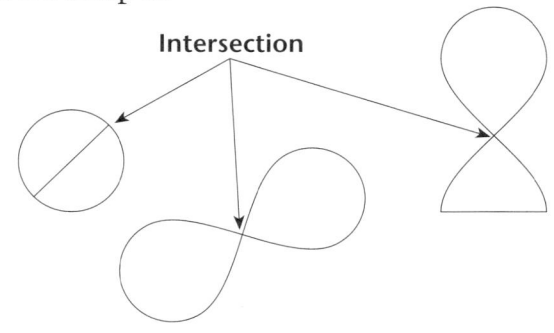
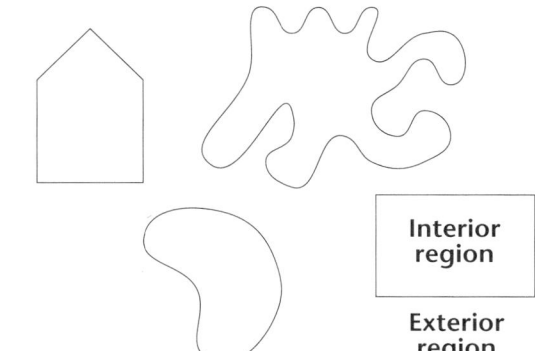

A **simple closed curve** (or Jordan curve) can be distorted into a circle and has **no intersections**. It divides an area into **only two regions** as shown on the left above.

Color the closed curves below red and the simple closed curves below yellow.

A. 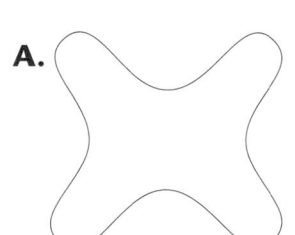   **B.** 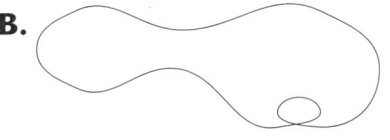   **C.**

D.

E.    **F.** 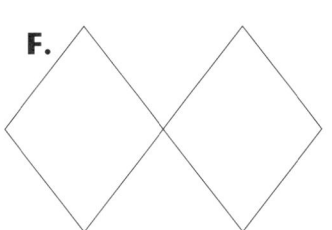   **G.** 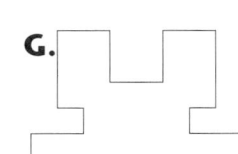   **H.**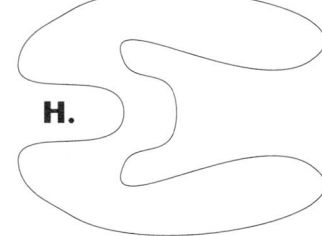

Using your imagination draw a pattern or animal which has twenty closed curves and twenty simple closed curves. Color the two types of curves in different colors.

World Teachers Press — Problem Solving Through Investigation – Book 1

The Circle

A *circle* is a closed curve with all points at the same distance from the center.

Parts of the circle

Radius: A straight line from the center to a point on the circumference.

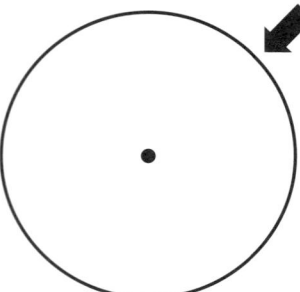

Circumference: The perimeter of the circle.

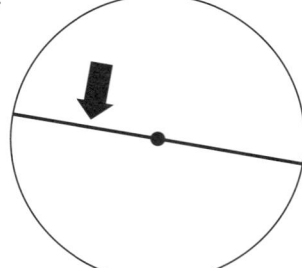

Diameter: A straight line joining two points of a circle and passing through the center.

Arc: Part of the circumference of a circle.

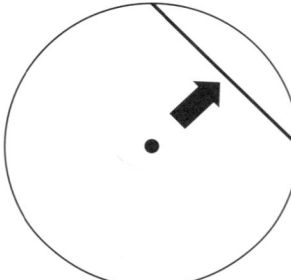

Chord: A straight line connecting the end points of an arc.

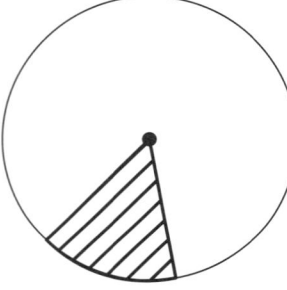

Sector: The area enclosed by two radii and an arc.

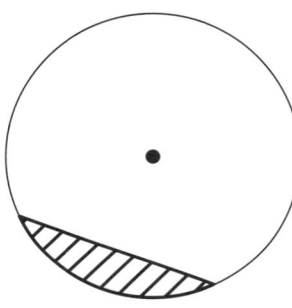

Segment: The area enclosed by an arc and a chord.

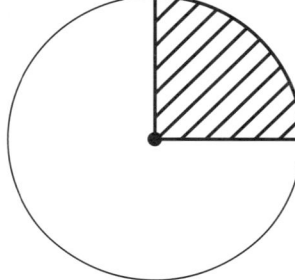

Quadrant: A quarter of a circle enclosed by two radii and an arc and containing a right angle.

Create patterns below by following the instructions and using the definitions above.

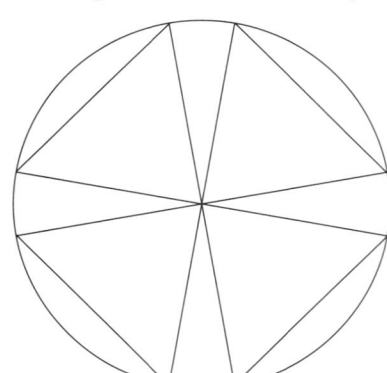

Pattern 1
Sectors – Blue
Segments – Yellow
Arcs in Sectors – Black
Chords – Red

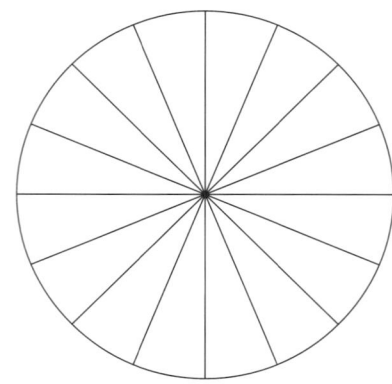

Pattern 2
Diameters – Red
Alternate Sectors – Yellow
Arcs in uncolored sectors – Black

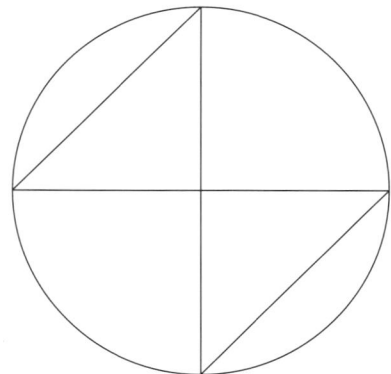

Pattern 3
Quadrants – Yellow
Radii – Brown
Circumference – Black
Segments – Green

Coordinates (1)

Coordinates are used in math and on maps to locate points by using ordered pairs of numbers as a reference. Look at the example on the right.

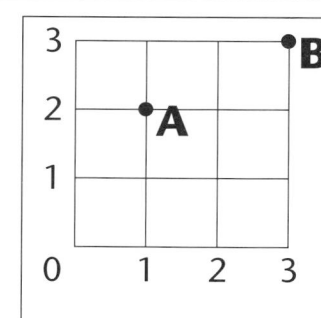

We always read along the horizontal axis first.
Point A is referenced by the ordered pair (1,2).
Point B is referenced by the ordered pair (3,3).

Complete the message below by finding the grid letter that goes with each ordered pair.

Message

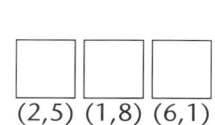

S	T	A	N	D		O	N
(4,5) (2,3) (5,2) (7,5) (6,7)   (2,8) (8,1)

Y	O	U	R
(8,8) (4,3) (1,1) (2,6)

C	H	A	I	R		A	N	D
(4,8) (3,1) (7,7) (1,4) (7,3)   (2,5) (1,8) (6,1)

S	A	Y		I
(8,4) (3,7) (3,2)   (8,6)

L	I	K	E
(5,6) (3,6) (6,4) (5,8)

M	A	T	H
(3,4) (8,3) (2,1) (1,6)

8	N	O	B	C	E	T	L	Y
7	S	V	A	P	E	D	A	Z
6	H	R	I	M	L	T	O	I
5	X	A	T	S	B	C	N	D
4	I	H	M	J	S	K	P	S
3	S	T	W	O	P	T	R	A
2	E	V	Y	L	A	X	T	P
1	U	T	H	A	F	D	P	N
	1	2	3	4	5	6	7	8

Now make up a message below for your partner by using the letters on the grid.

Coordinates (2)

Use ordered pairs (for example: 1,3; 2,4) to locate the correct points. Then enlarge each point you locate and join these enlarged dots with your ruler to outline a two-dimensional shape. Label the shape from the names given below. Always read the horizontal axis number first.

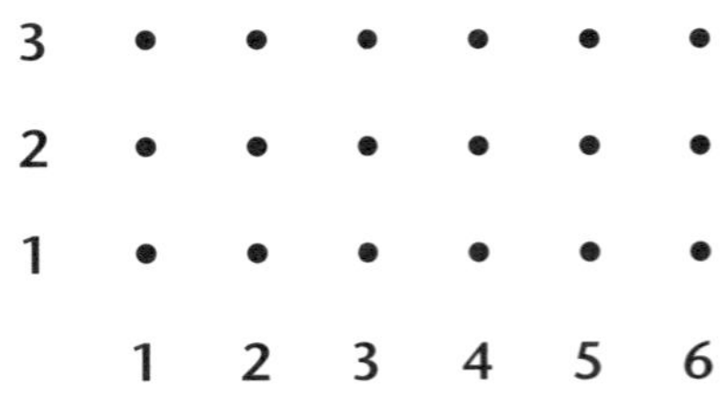

Pairs (1,1) (3,2) (5,3) (6,1) (3,1)

Name

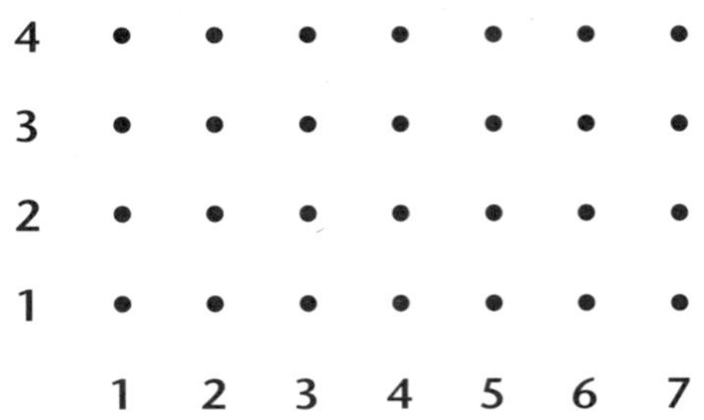

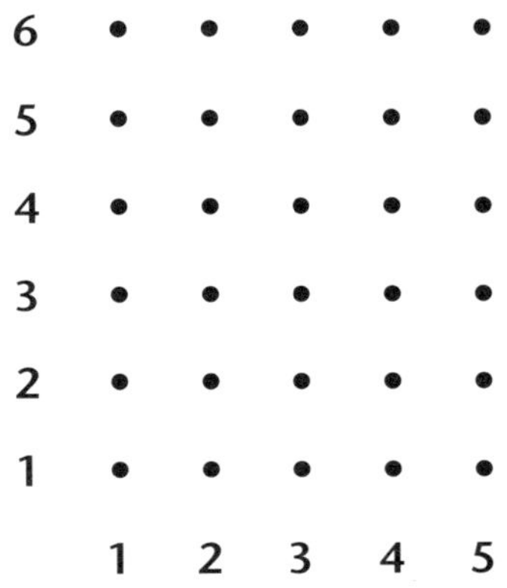

Pairs (2,1) (1,4) (3,6) (5,4) (4,1)

Name

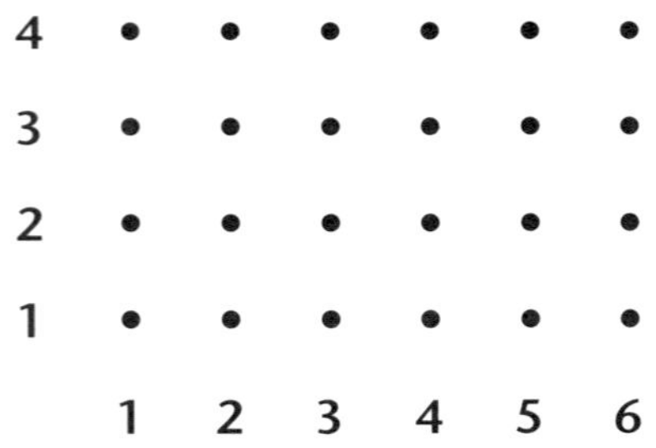

Pairs (1,2) (3,4) (5,4) (7,2)

Name

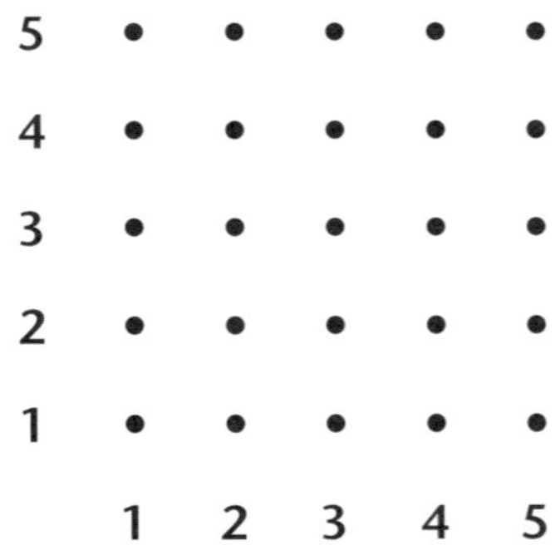

Pairs (1,1) (3,1) (4,4) (6,4)

Name

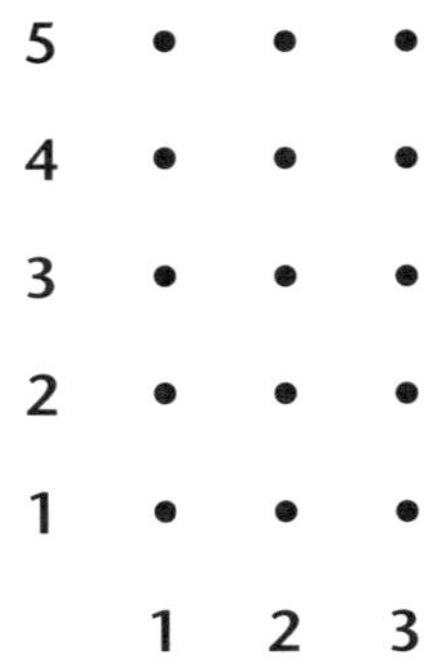

Pairs (1,1) (2,3) (3,5) (4,3) (5,1) (3,1)

Name

Pairs (2,1) (1,3) (3,3) (2,5)

Name

Shapes: Rhombus, Pentagon, Trapezium, Parallelogram, Scalene Δ, Isosceles Δ

A dictionary will help!

Area

Find the area of the **shaded part** in each diagram. There is a short way of doing each one without doing too many long calculations. You may use your calculator.

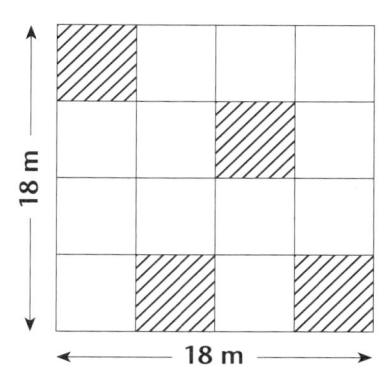

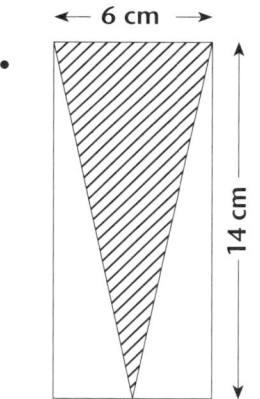

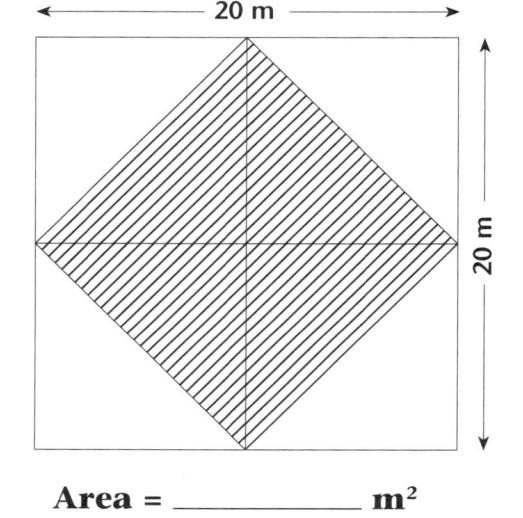

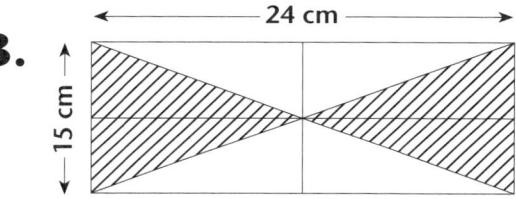

1. 16 cm × 16 cm square, with 8 cm marked. Area = _____ cm²

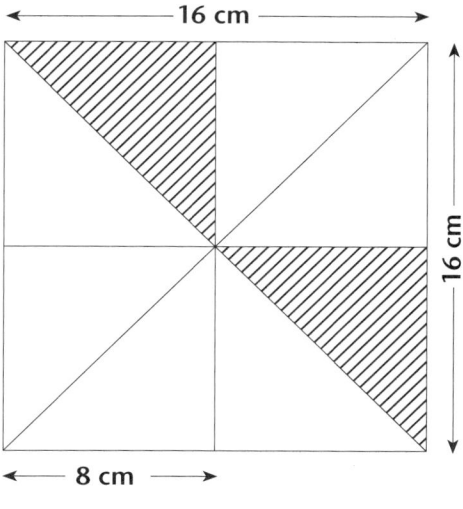

2. 18 m × 18 m square divided into grid with some shaded squares. Area = _____ m²

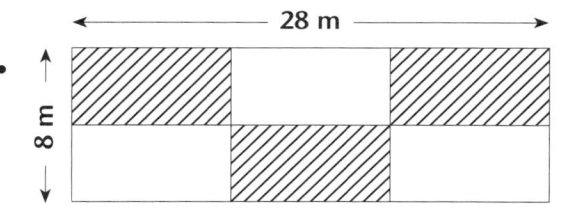

3. 24 cm × 15 cm rectangle with bowtie shading. Area = _____ cm²

4. Triangle with top 6 cm, height 14 cm. Area = _____ cm²

5. 28 m × 8 m rectangle with alternating shaded sections. Area = _____ m²

6. 20 m × 20 m square with inscribed diamond (rhombus) shaded. Area = _____ m²

7. Circle divided into 8 sections, one shaded. Area of circle is 264 cm². Area = _____ cm²

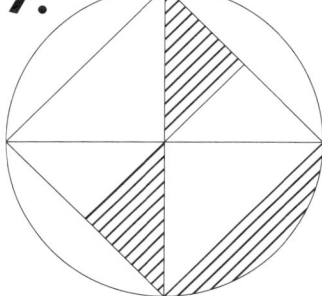

World Teachers Press — Problem Solving Through Investigation – Book 1

Line Segments

A line segment is a set of points with two end points.

With one point we cannot draw a line segment.

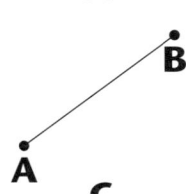

With two points we can draw one line segment.

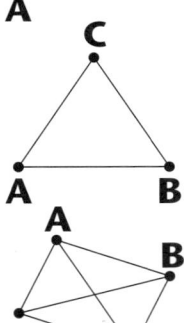

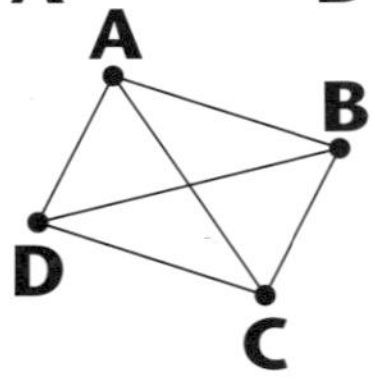

With three points we can draw three line segments.

With four points we can draw six line segments.

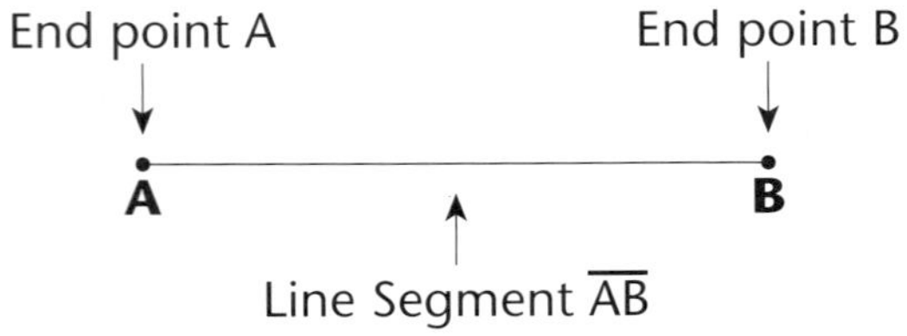

Line Segment $\overline{AB}$

	Points	Line segments	Shape outline
A	1	0	-
	2	1	-
	3	3	Triangle
	4	6	Quadrilateral
	5		
	6		
	7		
	8		
B	9		Nonagon
	10		Decagon
	11		Hendecagon
	12		Dodecagon

1. Complete the diagrams below and complete part A of the table to the right.

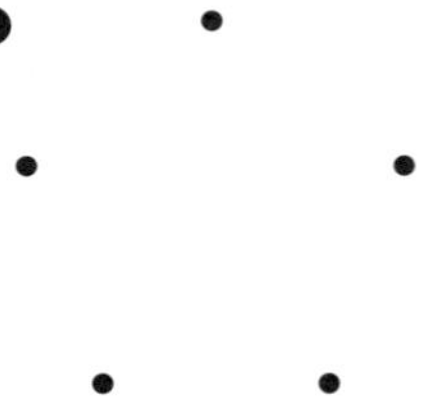

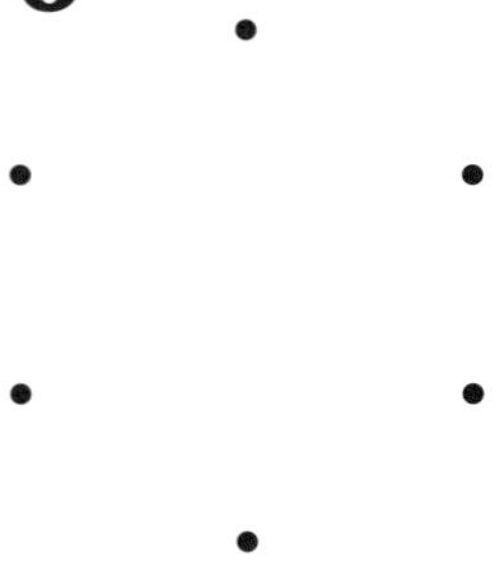

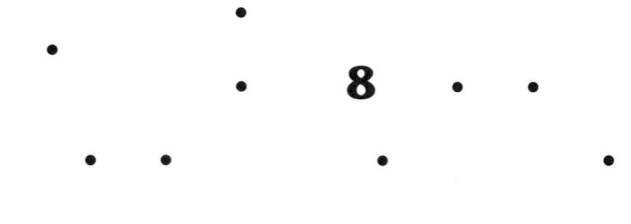

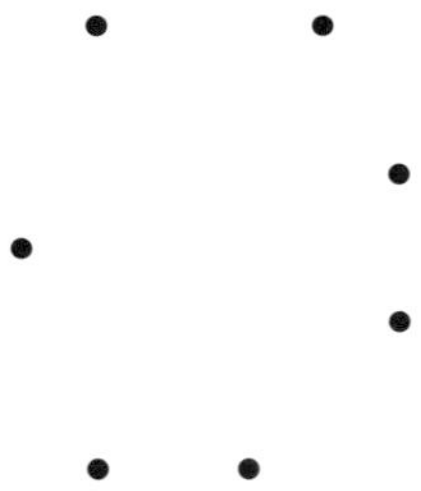

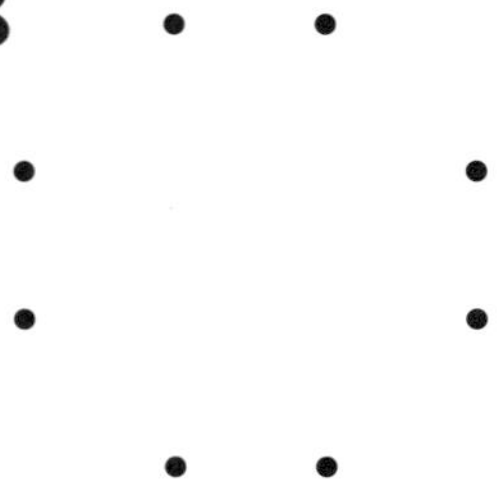

2. Look for the pattern down the second column of the table above.
   Then without drawing any diagrams complete the second column in Part B of the table.

Coordinates (3)

Coordinates are used on maps to locate places by using ordered pairs of numbers as a reference.

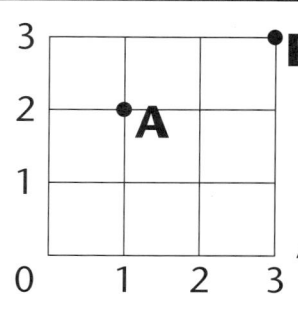

Point A is referenced by the ordered pair (1, 2).
Point B is referenced by the ordered pair (3, 3).

Always read along the horizontal axis first.

This system of coordinates was invented by the 17th century mathematician **Descartes** and is called **Cartesian coordinates** after him.

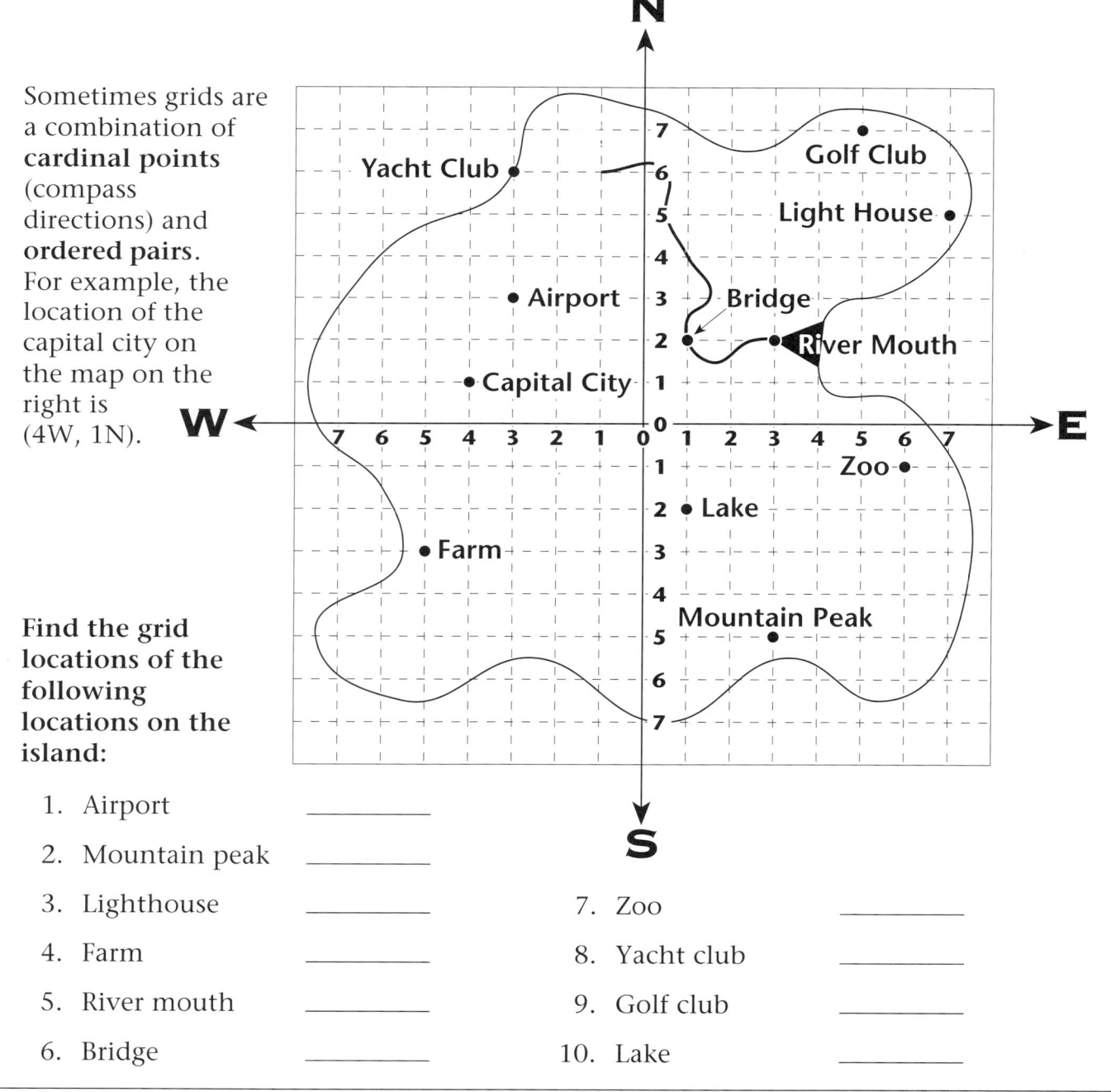

Sometimes grids are a combination of **cardinal points** (compass directions) and **ordered pairs**. For example, the location of the capital city on the map on the right is (4W, 1N).

Find the grid locations of the following locations on the island:

1. Airport _____
2. Mountain peak _____
3. Lighthouse _____
4. Farm _____
5. River mouth _____
6. Bridge _____
7. Zoo _____
8. Yacht club _____
9. Golf club _____
10. Lake _____

World Teachers Press — Problem Solving Through Investigation – Book 1

Rigid Shapes

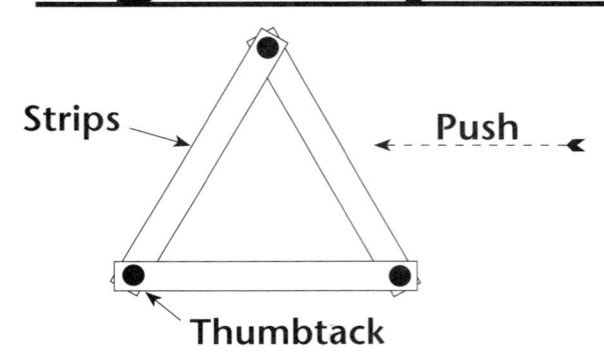

Strips / Push / Thumbtack

Make an equilateral triangle out of three equal strips of thick card and three thumbtacks. Push the triangle in the direction of the arrow. You will find it will not move because **a triangle is a rigid (strong) shape**.

Now do the same with a rectangle as shown. Push it. Is it rigid? _____

Put in a diagonal strut made of card, as shown in the diagram.

Does this make it rigid? _____

Why is this? _____

_____

_____

Thumbtack / Diagonal Strut / Push

Make the necessary shapes and complete the final two columns. The lengths of the sides of the shapes need not be equal. Find pictures of bridges, buildings and objects which clearly show the use of the two common rigid shapes (triangles, circles).

Shape	Number of sides	Number of struts to make rigid	Number of triangles formed
Triangle	3	0	1
Rectangle	4	1	2
Pentagon	5		
Hexagon	6		
Heptagon	7		
Octagon	8		

World Teachers Press — Problem Solving Through Investigation – Book 1

Answers

Page 6: Factors

Factors of 5 = 1, 5.
Factors of 21 = 1, 3, 7, 21.
Factors of 13 = 1, 13.
Factors of 36 = 1, 2, 3, 4, 6, 9, 12, 18, 36.
Factors of 40 = 1, 2, 4, 5, 8, 10, 20, 40.
Factors of 27 = 1, 3, 9, 27.
Factors of 19 = 1, 19.

Page 7: Factor Patterns

Factors of 12 = 1, 2, 3, 4, 6, 12.
Factors of 20 = 1, 2, 4, 5, 10, 20.

Page 8: Products and Isometric Shapes

Shape A: trapezium
Shape B: square
Shape C: parallelogram
Shape D: pentagon

Page 9: Odd Numbers

1. Answers will vary.
2. 17, 89, 375, 841, 1253
   Odd numbers can be picked by last digit.
3.

x	1	3	5	7
11	11	33	55	77
9	9	27	45	63
3	3	9	15	21
13	13	39	65	91

   The statement is true.
4. (b) 63 (odd)   (c) 495 (odd)
   (d) 105 (odd) (e) 35 (odd)
   (f) 27 (odd)   (g) 125 (odd)
   (h) 45 (odd)

Page 10: Prime Numbers

A. prism
B. array
C. tally
D. radii

Page 11: Composite Numbers

Shape A is a rectangular prism and has 6 faces, 8 vertices and 12 edges.
Shape B is a triangular prism and has 5 faces, 6 vertices and 9 edges.

Page 12: Prime and Composite Numbers

1. An even number has 2 as one of its factors.
2. A prime number has two factors: 1 and itself.
3. Answers will vary. All even numbers greater than 2 may be found by adding together two prime numbers.

Page 13: Commutative Law

1.

+	3	4	5	6
3	6	7	8	9
4	7	8	9	10
5	8	9	10	11
6	9	10	11	12

x	3	4	5	6
3	9	12	15	18
4	12	16	20	24
5	15	20	25	30
6	18	24	30	36

2. Answers will vary.

Page 14: Square and Triangular Numbers

1. Square Numbers: 3, 5, 7, 9, …
   Triangular Numbers: 3, 4, 5, 6, …
2. 16 = 6 + 10
   25 = 10 + 15
   36 = 15 + 21
3. Answers will vary.

Page 15: Probability

Answers will vary. The number of heads and tails can't be predicted as it is a chance process.

Page 16: Magic Squares

1.

4	9	2
3	5	7
8	1	6

2.

4	6	5
6	5	4
5	4	6

3.

16	3	2	13
5	10	11	8
9	6	7	12
4	15	14	1

(a) 34   (b) 34, 34, 34, 34

World Teachers Press — Problem Solving Through Investigation – Book 1

Answers (continued)

Page 17: Fun with the Magical 9!
1. The numbers above and below the boxes are the same.
2. There are no remainders.
3. All answers contain the same numerals (e.g. 66, 33). There are no remainders.

Page 18: Operations – Calculator Work
1. lob   2. gill   3. bible
4. sleigh   5. log   6. sole
7. loss   8. shoe   9. hose
10. bill   11. hobble 12. bog
13. lobe   14. lie   15. geese

Page 19: Number Patterns (1)
Answers will vary.

Page 20: Number Patterns (2)
289, 361, 441, 529, 625, 729, 841.
The boxed numbers are square numbers.

Page 21: Number Patterns (3)
1. 87, 78, 89   2. 37, 28, 17
3. 37, 47, 46   4. 6, 17, 8
5. 67, 87, 89   6. 46, 37, 27
7. 56, 67, 78   8. 86, 77, 88
9. 17, 8, 9

Page 22: Number Systems
Egyptian
1. 100,011   2. 123
3. 11,202   4. 1,024
Roman
1. 1,600   2. 57
3. 94   4. 25
Babylonians
1. 72   2. 30
3. 23   4. 90
Mayan
1. 17   2. 28
3. 11   4. 24

Page 23: Index Notation – Calculator Work
1. his   2. hole   3. beg
4. bee   5. shell   6. ebb
7. bliss   8. soil   9. hoe
10. gloss   11. oil   12. boil
13. ogle   14. bile   15. hobbies

Page 24: Mapping
The relation sign is given its name because it shows the relationship between the sets.
1. Answers will vary.
2. Ottawa, Spain, Greece.
3. Answers will vary.

Page 25: Perimeter
The perimeter of the robot is 58 cm.
The perimeter of the tortoise is 45 cm.

Page 26: Perimeter and Area
1. All houses have a perimeter of 36 m.
2. A = 56 m^2
   B = 72 m^2
   C = 81 m^2
   D = 80 m^2
   E = 66 m^2
3. House C.
4. Answers will vary.

Page 27: Triangles
3. Isosceles triangles.
4. congruent
5. 60

1. The base angles are equal.
2. The base angles in each new triangle are equal.
3. Yes.

Page 28: Angles
A. 1. The length of a straight line extending from the center to the circumference of a circle.
   2. The length of a straight line passing from one side of a circle to the other, through its center.
   3. The length of the outer boundary of a simple closed curve or polygon.
B. 6. Angles are always 90°.
C. Yes.

Answers (continued)

Page 29: Decades and Centuries
1. 7 decades (1806 to 1876)
2. 6 centuries (1215 to 1815)
3. 5 decades (1903 to 1953)
4. 4 centuries (1588 to 1988)
5. 10 decades (1788 to 1888)
6. 2 centuries (1776 to 1976)

Page 30: Polygons
No answers required.

Page 31: Duplation
1. Answers will vary.
2. Answers set out as per example.

Page 32: Conservation of Area
Lower shapes S and T both fit into the upper shapes when cut and rearranged.

Page 33. Congruency
Shapes A, C and D are congruent.

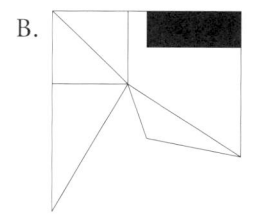

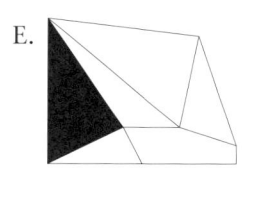

Page 34: Reflective Symmetry
Answers will vary. Coloring should follow instructions.

Page 35: Closed Curves
Curves a, d, g, h are yellow.
Curves b, c, e, f are red.

Page 36: The Circle
No answers required.

Page 37: Coordinates (1)
The message is: "Stand on your chair and say 'I like math'".

Page 38: Coordinates (2)
Left column first, top to bottom: scalene triangle, trapezium, isosceles triangle, pentagon, parallelogram, rhombus.

Page 39: Area
1. 64 cm²
2. 81 m²
3. 180 cm²
4. 42 cm²
5. 112 m²
6. 200 m²
7. 66 cm²

Page 40: Line Segments
Missing parts of the table are:
- 10 pentagon
- 15 hexagon
- 21 heptagon
- 28 octagon
- 36
- 45
- 55
- 66

Page 41: Coordinates (3)
1. 3W, 3N
2. 3E, 5S
3. 7E, 5N
4. 5W, 3S
5. 3E, 2N
6. 1E, 2N
7. 6E, 1S
8. 3W, 6N
9. 5E, 7N
10. 1E, 2S

Page 42: Rigid Shapes
The strut in the rectangle makes it rigid because it divides the rectangle into two strong triangles. The numbers in columns 3 and 4 follow consecutively.

About the Author

George Moore, has taught and lived in Perth, Western Australia, New South Wales (State), and the United Kingdom. George has been a practicing classroom teacher for over 30 years, with experience in primary and secondary areas. He has held promotional positions in England and Australia.